普通高等院校"十三五"规划教材

C++面向对象程序设计双语教程
（第2版）
Object-Oriented Programming in C++
(2nd Edition)

刘嘉敏　马广焜　常燕　朱世铁　编著

国防工业出版社
·北京·

图书在版编目(CIP)数据

C++面向对象程序设计双语教程/刘嘉敏等编著.
—2版.—北京：国防工业出版社，2015.8
ISBN 978 – 7 – 118 – 10365 – 6

Ⅰ.①C… Ⅱ.①刘… Ⅲ.①C语言 – 程序设计 – 双语教学 – 教材 Ⅳ.①TP312

中国版本图书馆 CIP 数据核字(2015)第 191135 号

※

国防工业出版社出版发行
（北京市海淀区紫竹院南路23号 邮政编码100048）
涿中印刷厂印刷
新华书店经售

*

开本 787×1092 1/16 印张 18 字数 405 千字
2015 年 8 月第 2 版第 1 次印刷 印数 1—2500 册 定价 36.00 元

（本书如有印装错误，我社负责调换）

国防书店：(010)88540777　　　　发行邮购：(010)88540776
发行传真：(010)88540755　　　　发行业务：(010)88540717

第 2 版前言

本书第 1 版自出版以来,在作者学校计算机科学与技术专业本科生教学已使用 2 年,得到了广大师生的赞许,而且也被其他院校用作程序设计初学者的双语教材。

本书从面向对象程序设计的特点和工程应用角度出发,递进式组织各个章节的知识点,采用易懂的实例,引导初学者进入面向对象程序设计之门。新版教材在前版的基础上,结合当今学生对新知识的认知程度,以保持原有的特色为前提,对前版教材进行了修改和补充,适当地增加了案例和练习题,使得初学者对相关概念和实现方法更加理解和掌握。例如,在第 3 章,增加了对面向对象封装性的认识和实例,以及 UML 图对类的描述,同时在第 3 章~第 8 章增加了案例和练习题。

虽然本教材以英语叙述,但在保持英文原汁原味的基础上,符合中国学生思维习惯和学习习惯,并针对主要的知识点,通过"教学目标""重点注释""实例""概念加注框""思考题""词汇小贴士""案例学习"和"练习题"组织形式,重点突出、通俗易懂,便于学生阅读和掌握。

面向对象程序设计是实践性较强的课程,因而上机实践是知识点学习和巩固必不可少的环节。为此本教材针对每个知识点配有完整代码和运行结果,使学生可以确定程序预期结果,通过输出结果与程序语句联系在一起,为学生提供实践和自学的方式。

本教材旨在培养学生学会面向对象程序设计的基本概念、基本思想和方法,虽然本教材以 C++程序设计语言来描述,但其不是 C++的面向对象程序设计,其内涵与用其他语言描述面向对象程序设计是一致的,因而本教材在许多概念上不过分强调 C++语言的细节,而着重对面向对象程序设计的基本概念、思想和方法的认识。

本书得到沈阳工业大学信息科学与工程学院、美国南加州大学王博涵同学的帮助,以及作者家人的支持,而且书中引用了其他同行的工作成果,在此一并表示衷心感谢。

由于作者英语水平有限,书中难免存在错漏和不妥之处,恳请读者提出宝贵意见。

编者
2015 年 6 月

第 1 版前言

面向对象程序设计(Object-Oriented Programming,OOP)是一种以对象为基础,以事件或消息来驱动对象执行相应处理的程序设计方法。它将对象作为程序的基本单元,将数据和对数据的操作封装在一起,以提高软件的重用性、灵活性和扩展性。它是一种为现实世界建立对象模型,利用继承、多态和消息机制设计软件的先进技术,非常适合于大型应用程序与系统程序的开发,是当今计算机程序设计倡导的主流技术。

为使读者快速建立面向对象的概念、循序渐进地理解 OOP 技术的各知识点内容、系统地建立 OOP 完整的知识架构及利用 OOP 方法完整地解决一个实际应用问题,本书以 OOP 知识节点递进式组成各章节内容,并以实际应用问题作为样例贯穿于全书。本书内容是作者根据十余年从事 OOP 课程教学、OOP 课程设计体会及多年指导学生毕业设计、课程设计和计算机程序竞赛所积累的经验而精心组织的,各章节内容结构清晰,便于学习和掌握。

在多种支持面向对象程序设计语言中,正是由于 C++ 语言以其与 C 语言兼容、高运行效率等优良特性,使大量 C 程序员通过 C++ 的帮助迅速掌握了面向对象的概念和方法,全面促进了面向对象技术的应用,从而使 C++ 成为最有影响、最广泛应用的面向对象程序设计语言。因此,本书针对掌握 C 语言程序设计方法的初学者,以 C++ 语言为基础讲述面向对象程序设计概念和实现方法。

以英文撰写,旨在满足高校提倡的外语教学、双语教学需求,提高学生专业英语阅读能力。由于目前国内鲜见由国内学者自编、符合中国学生思维习惯和学习习惯的 OOP 英文教材,因此作者倾注了多年心血奉献此书。该书作为校内教材在作者学校计算机科学与技术专业本科生教学已使用 4 年,曾修订过两版,得到了广大师生的赞许。

在内容组织结构上,为学生提供有:

"教学目标(Objectives)",告诉学生每章重点知识,让学生学完一章后判定是否达到这些目标,帮助学生建立学习信心。

"重点注释(Notes)",对重点知识和易混淆知识点加以提示和强调,告诉学生必须清楚这些知识点,并且知道如何正确使用。

"实例(Examples)",对主要知识点给出通俗易懂、适合初学者的实例,实例都配有完整的代码和运行结果,使学生可以确定程序预期结果,通过输出结果与程序语句联系在一起,为学生提供学习和巩固概念的方法。

"概念加注框",将各章节涉及基本概念详解归纳集聚在矩形边框内,供学生一目了然地阅读、学习和掌握。

"思考题(Think Over)",帮助学生回顾和总结重点知识,检查对知识的学习掌握程度。

"词汇小贴士(Word Tips)",按字母顺序列出各章节中的生词,给出词性和专业的解释,帮助学生快速、准确理解OOP专业知识,提高外语阅读能力。

"练习题(Exercises)",提供从概念理解、基本语句运用到简单问题求解的各种类型练习题,便于学生对知识的理解和运用。习题的难度适宜,学生基本能独立完成。较复杂问题和团队合作大型综合训练在课程实验和课程设计中安排。

本书共分8章,每章的主要内容有:

第1章 Introduction,介绍了什么是程序设计,如何面对实际问题设计相应的程序代码的步骤;简要地介绍了从机器语言到高级语言的程序设计发展历史;介绍了两种主流的程序设计方法,着重介绍了面向对象程序设计的概念和基本特点;讨论了C++程序设计语言的历史和C++语言的学习方法。

第2章 Basic Facilities,讨论了与C语言不同的C++的基础知识,以结构化程序设计方法,介绍了基本的输入/输出流、常量、函数、引用和命名空间,可以使学生顺利地从C语言过渡到C++语言的程序设计。

第3章 Classes and Objects(Ⅰ),从这章开始学习面向对象程序设计的基本概念和方法,本章为类与对象设计的基础部分,介绍了数据抽象和信息隐藏的概念,详细讲述了抽象数据类型(ADT)类的基本设计方法、类中主要成员构造函数、析构函数的定义和对象的生成。

第4章 Classes and Objects(Ⅱ),是进阶使用对象和类的部分,讨论了类中的一些特殊成员的定义和实现,如静态成员、常量成员、对象成员、拷贝成员和友元,以及各种类型的对象声明和使用。

第5章 Operator Overloading,介绍了如何对用户定义数据类型使用现有运算符进行运算,主要介绍了运算符重载的基础、运算符的限制、重载成员函数和非成员函数、一元和二元运算符以及类型转换。

第6章 Inheritance,继承是软件重用的一种形式,介绍了面向对象程序设计的主要特性之一———继承概念,重点介绍了派生类的成员函数、访问控制、构造函数和析构函数的定义,讨论了多继承的模糊性,以及如何运用虚基类避免模糊性的设计。

第7章 Polymorphism and Virtual Functions,多态是面向对象程序设计的另一个主要特性,从同一基类继承多个类和设计相同的接口的方法,使系统更易于扩展。本章介绍了两种多态的特点、虚函数的工作原理和设计方法,并介绍了抽象基类的实现过程。

第 8 章 Templates,模板也是软件重用的一种形式,它使程序员可以利用参数形式描述抽象数据类型的本质,然后用少量代码生成特定的抽象数据类型。本章介绍了函数模板和类模板的定义和实现方法。

书中所有源代码在 Microsoft Visual C++ 6.0 环境中调试和运行通过。

该教材适合于 40~54 学时教学,配有 PPT 教学课件,并提供实验指导书、实验题目详解代码、OOP 课程设计题目,有需要的任课教师请垂询 jmliu@sut.edu.com。

本书由刘嘉敏统编,主要编写了第 1、2、3、4、5 章,参编了第 7 章;马广焜编写了第 6、7、8 章;常燕参编了第 1 章和练习题,并且完成了全书的排版和校对;朱世铁参编了练习题和实验指导书。英国贝德福德大学 Des Stephens 老师利用他的业余时间,非常耐心地修正了书中英语错误,借此对他的支持和帮助致以衷心的感谢。另外,沈阳工业大学的杨丹、邱辉、靳长旭、闫博、郭云龙、赵天育、王爱新和张荣铭等同学协助完成每章的词汇表,在此致谢。此外,书中引用了其他同行的工作成果,在此一并表示衷心感谢。

由于作者英语水平有限,书中难免会有错漏和不妥之处,恳请读者提出宝贵意见,在此衷心地表示感谢。

编者
2012 年 11 月

目 录

Chapter 1　Introduction ··· 1
　1.1　Overview of Programming ··· 1
　　1.1.1　What Is Programming? ··· 1
　　1.1.2　How Do We Write a Program? ·· 3
　1.2　The Evolution of Programming Language ································ 5
　　1.2.1　Assembly and Machine Languages ···································· 5
　　1.2.2　Early Languages ·· 6
　　1.2.3　Later-Generation Languages ·· 7
　　1.2.4　Modern Languages ·· 7
　1.3　Programming Methodologies ·· 8
　　1.3.1　Structured Programming ··· 8
　　1.3.2　Object-Oriented Programming ·· 10
　1.4　Object-Oriented Programming ·· 12
　1.5　C++ Programming Language ·· 15
　　1.5.1　History of C and C++ ··· 15
　　1.5.2　Learning C++ ·· 16
　Word Tips ·· 17
　Exercises ··· 18

Chapter 2　Basic Facilities ·· 19
　2.1　C++ Program Structure ·· 19
　2.2　Input / Output Streams ·· 21
　2.3　Constant ·· 22
　2.4　Functions ··· 24
　　2.4.1　Function Declarations ·· 24
　　2.4.2　Function Definitions ·· 25
　　2.4.3　Default Arguments ·· 26
　　2.4.4　Inline Functions ·· 28
　　2.4.5　Overloading Functions ·· 29

2.5 References ·········· 33
   2.5.1 Reference Definition ·········· 33
   2.5.2 Reference Variables as Parameters ·········· 37
   2.5.3 References as Value-Returning ·········· 38
   2.5.4 References as Left-Hand Values ·········· 40
2.6 Namespaces ·········· 41
Word Tips ·········· 45
Exercises ·········· 45

Chapter 3 Classes and Objects (I) ·········· 49

3.1 Structures ·········· 49
   3.1.1 Defining a Structure ·········· 49
   3.1.2 Accessing Members of Structures ·········· 50
   3.1.3 Structures with Member Functions ·········· 52
3.2 Data Abstraction and Classes ·········· 53
   3.2.1 Data Abstraction ·········· 53
   3.2.2 Defining Classes ·········· 54
   3.2.3 Defining Objects ·········· 55
   3.2.4 Accessing Member Functions ·········· 55
   3.2.5 In-Class Member Function Definition ·········· 58
   3.2.6 File Structure of an Abstract Data Type ·········· 60
3.3 Information Hiding ·········· 62
3.4 Access Control ·········· 63
3.5 Constructors ·········· 65
   3.5.1 Overloading Constructors ·········· 66
   3.5.2 Constructors with Default Parameters ·········· 67
3.6 Destructors ·········· 70
   3.6.1 Definition of Destructors ·········· 70
   3.6.2 Order of Constructor and Destructor Calls ·········· 71
3.7 Encapsulation ·········· 74
3.8 Case Study: A GradeBook Class ·········· 75
Word Tips ·········· 78
Exercises ·········· 79

Chapter 4 Classes and Objects (II) ·········· 82

4.1 Constant Member Functions and Constant Objects ·········· 82

4.2 *this* Pointers 84
4.3 Static Members 86
 4.3.1 Static Data Members 88
 4.3.2 Static Member Functions 91
4.4 Free Store 92
4.5 Objects as Members of Classes 96
4.6 Copy Members 102
 4.6.1 Definition of Copy Constructors 102
 4.6.2 Shallow Copy and Deep Copy 105
4.7 Arrays of Objects 114
 4.7.1 Initialize an Array of Objects by Using a Default Constructor 114
 4.7.2 Initialize an Array of Objects by Using Constructors with Parameters 117
4.8 Friends 117
 4.8.1 Friend Functions 118
 4.8.2 Friend Classes 121
4.9 Case Study: Examples of Used-defined Types 122
 4.9.1 A Better Date Class 122
 4.9.2 A GradeBook Class with Objects of the Student Class 128
Word Tips 134
Exercises 134

Chapter 5 Operator Overloading 140

5.1 Why Operator Overloading Is Need 140
5.2 Operator Functions 141
 5.2.1 Overloaded Operators 141
 5.2.2 Operator Functions 141
5.3 Binary and Unary Operators 145
 5.3.1 Overloading Binary Operators 145
 5.3.2 Overloading Unary Operators 146
5.4 Overloading Combinatorial Operators 150
5.5 Mixed Arithmetic of User-Defined Types 154
5.6 Type Conversion of User-Defined Types 154
5.7 Examples of Operator Overloading 156
 5.7.1 A *Complex Number* Class 156
 5.7.2 A *String* Class 164
Word Tips 170

Exercises ··· 170

Chapter 6 Inheritance ·· 174

6.1　Class Hierarchies ··· 174

6.2　Derived Classes ·· 175

　　6.2.1　Declaration of Derived Classes ··· 175

　　6.2.2　Structure of Derived Classes ·· 176

6.3　Constructors and Destructors of Derived Classes ····························· 179

　　6.3.1　Constructors of Derived Classes ··· 179

　　6.3.2　Destructors of Derived Classes ·· 182

　　6.3.3　Order of Calling Class Objects ··· 183

　　6.3.4　Inheritance and Composition ·· 186

6.4　Member Functions of Derived Classes ·· 186

6.5　Access Control ··· 189

　　6.5.1　Access Control in Classes ·· 189

　　6.5.2　Access to Base Classes ·· 191

6.6　Multiple Inheritance ·· 195

　　6.6.1　Declaration of Multiple Inheritance ······································· 196

　　6.6.2　Constructors of Multiple Inheritance ····································· 198

6.7　Virtual Inheritance ··· 199

　　6.7.1　Multiple Inheritance Ambiguities ··· 199

　　6.7.2　Trying to Solve Inheritance Ambiguities ································· 200

　　6.7.3　Virtual Base Classes ·· 203

　　6.7.4　Constructing Objects of Multiple Inheritance ·························· 205

Word Tips ·· 207

Exercises ··· 208

Chapter 7 Polymorphism and Virtual Functions ································ 218

7.1　Polymorphism ·· 218

　　7.1.1　Concept of Polymorphism ·· 218

　　7.1.2　Binding ·· 219

7.2　Virtual Functions ··· 221

　　7.2.1　Definition of Vitual Functions ·· 222

　　7.2.2　Extensibility ··· 225

　　7.2.3　Principle of Virtual Functions ··· 228

　　7.2.4　Virtual Destructors ·· 229

  7.2.5 Function Overloading and Function Overriding ·················· 230
 7.3 Abstract Base Classes ··· 233
 7.4 Case Study: A Mini System ··· 236
 Word Tips ·· 241
 Exercises ··· 242

Chapter 8 Templates ··· 248

 8.1 Template Mechanism ··· 248
 8.2 Function Templates and Template Functions ······················ 249
  8.2.1 Why We Use Function Templates? ························· 249
  8.2.2 Definition of Function Templates ··························· 250
  8.2.3 Function Template Instantiation ···························· 251
  8.2.4 Function Template with Different Parameter Types ······ 253
  8.2.5 Function Template Overloading ····························· 254
 8.3 Class Templates and Template Classes ····························· 255
  8.3.1 Definition of Class Templates ································ 255
  8.3.2 Class Template Instantiation ································· 258
 8.4 Non-Type Parameters for Templates ································ 261
 8.5 Derivation and Class Templates ····································· 262
 8.6 Case Study: An Example of the Vector Class Template ········ 264
 Word Tips ·· 270
 Exercises ··· 270

References ·· 272

Chapter 1
Introduction

> *High thoughts must have high language.*
> —*Aristophanes*

Objectives
- To understand what a computer program is
- To be able to list the basic stages involved in writing a computer program
- To recognize two programming methodologies
- To know about characteristics of object-oriented programming

1.1 Overview of Programming

1.1.1 What Is Programming?

Much of human behavior and thought is characterized by logical sequences. Since infancy, you have been learning how to act, how to do things. And you have learned to expect certain behavior from other people.

On a broad scale, mathematics never could have been developed without logical sequences of steps for solving problems and proving theorems. Mass production never would have worked without operations taking place in a certain order. Our whole civilization is based on the order of things and actions. We create order, both consciously and unconsciously, through a process we call **programming**.

Programming is planning how to solve a problem. No matter what method is used—pencil and paper, slide rule, adding machine, or computer—problem solving requires programming. Of course, how one program depends on the device one uses in problem solving.

Programming is planning the performance of a task or an event.
Computer is a programmable device that can store, retrieve, and process data.
Computer programming is the process of planning a sequence of steps for a computer to follow.
Computer program is a sequence of instructions to the performed by a computer.

Back to programming—"planning how to solve a problem", note that we are not actually solving a problem. The computer is going to do that for us. If we could solve the problem ourselves we would have no need to write the program. The premise for a program is that we don't have the time, tenacity or memory capabilities to solve a problem but we do know how to solve it so can instruct a computer to do it for us.

A simple example of this is what is the sum of all integers from 1-10,000. If you wanted to you could sit down with a pencil and paper or a calculator and work this out however the time involved plus the likelihood that at some point you would make a mistake makes that an undesirable option. However you can write and run a program to calculate this sum in less than 5 minutes.

Example 1-1: An example of programming.

```
/* File:example1_1.c
 * The program calculates the sum of all integers from 1 to 10,000.
 */
1    #include "stdio.h"
2    #define MAX 10000
3
4    int main()
5    {
6        long sum = 0, number;
7
8        for( number = 1; number <= MAX; number ++)
9        {
10           sum += number;
11       }
12
13       printf("The sum of all integers from 1 to %ld is : %ld\n", MAX, sum);
14       return 0;
15   }
16
```

This example gives the result 50,005,000. As it happens you can verify this because you know that the sum of integers from 1-N can be calculated as

$(N+1)*(N/2)$

$(10000 + 1)*(10000/2) = 10001*5000 = 50005000$

So you have solved the problem of how to calculate the sum of all integers from 1 – 10,000 and the computer has solve the problem of calculating the sum of all integers from 1-10,000.

The computer allows us to do tasks more efficiently, quickly, and accurately than we could by hand—if we could do by hand at all. In order to use this powerful tool, we must specify what we want done and the order in which we want it done. We do this through programming.

However, we have to know the major differences between humans and computers.

Humans have judgment and free will and will not run any instruction they deem not required or nonsensical, whereas a computer will do exactly what it is told with no judgment on the need or sanity of the instruction.

1.1.2 How Do We Write a Program?

To write a sequence of instructions for a computer to follow, we must go through a two-phase process: problem solving and implementation (see Figure 1-1).

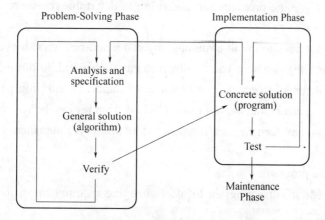

Figure1-1 Programming process

Problem Solving Phase

1. *Analysis and specification.* Understand (define) the problem and what the solution must do.

Every program starts with a specification. This may be a several hundred page document from your latest client or one small paragraph from your professor and pretty much anything in-between.

The specification is very important and specification writing is a whole sub-branch of programming. The important thing is that the specification is correct otherwise, to use a well-known computing adage, garbage in garbage out.

However, specifications are rarely perfect and it is not at all uncommon for the specification to go through several iterations before being finally agreed on.

2. *General solution (algorithm).* Develop a logical sequence of steps that solves the problem.

Having found out how to solve the problem you can then jump in and start coding, right? Wrong. Now is the time to do some program design. Now, how much design is required, and where the design is stored, is rather dependent on the complexity of the program, the experience of the user and the purpose of a program. For instance a simple program being written by an experienced programmer just as a temporary project tool (i.e. will only be in use for a day or 2) probably only requires a little thought about the program design.

The design, once you have it, is not set in stone. It just gives the current ideas about how the program will be written. I do not think I have seen a project where the final code exactly matches the initial design. Also as the specification changes so will the design. But having it will give a clear place to start coding from and will ultimately lead to better more maintainable code.

3. *Verify*. Follow the steps exactly to see if the solution really does solve the problem.

Implementation Phase

1. *Concrete solution (program)*. Translate the algorithm into a programming language.

2. *Test*. Have the computer follow the instructions. Then manually check the results. If you find errors, analyze the program and algorithm to determine the source of the errors, and then make corrections.

Testing does not have to involve running any code. Source review by your peers is a good form of testing too. This is where you sit down round a table and go through the code line by line and examine it for logic errors, conformance to standards and programming errors. This can actually throw up errors that are not made obvious from testing the software by running it.

Once a program has been written, it enters a third phase: maintenance.

Maintenance Phase

1. *Use*. Use the program.

2. *Maintain*. Modify the program to meet changing requirement or to correct any errors that show up in using it.

The programmer begins the programming process by analyzing the problem and developing a general solution called an algorithm. Understanding and analyzing a problem take up much more time than Figure 1-1 implies. They are the heart of the programming process.

> **Algorithm** is a step-by-step procedure for solving a problem in a finite amount of time.

An algorithm is verbal or written description of a logical sequence of actions (or events). After developing a general solution, the programmer tests the algorithm, walking through each step mentally or manually. If the algorithm does not work, the programmer repeats the problem-solving process, analyzing the problem again and coming up with another algorithm.

When the programmer is satisfied with the algorithm, he/she translates it into a programming language. Although a programming language is simple in form, it is not always easy to use. Programming forces you to write very simple, exact instruction. Translating an algorithm into a programming language is called *coding the algorithm*.

Once a program has been put into use, it is often necessary to modify it. Modification may involve fixing an error that is discovered during the use of the program or changing the program in response to changes in the user's requirements. Each time the program is modified, it is necessary to repeat the problem-solving and implementation phase for those aspects of the program that change. This phase of the programming process is known as maintenance and actually accounts for the majority of the effort expended on most programs. Together, the problem-solving, implementation and maintenance phases constitute the *program's life cycle*.

1.2 The Evolution of Programming Language

"As long as there were no machines, programming was no problem at all; when we had a few weak computers, programming became a mild problem and now that we have gigantic computers, programming has become an equally gigantic problem. In this sense the electronic industry has not solved a single problem, it has only created them — it has created the problem of using its product" (E.W. Dijkstra, Turing Award Lecture, 1972).

Nowadays, there are many programming languages, for example, C, C++, Ada, Pascal, Prolog, FORTRAN, Modula3, Lisp, Java, Scheme. This alphabet soup is the secret power of modern software engineering. As high level computer programming languages, they provide enormous flexibility and abstraction. Programmers are separated from the physical machine allowing complex problem solutions without fretting with the difficulties of ones and zeros. This idea is clarified by an analogy. If we had to think about every phonetic sound made while speaking, communication of abstract ideas would be next to impossible. Much the same way, programming directly with ones and zeros would focus the designer's attention on trivial hardware details instead of on designing abstract solutions. Considering the historical trend that created high level programming, we believe that certain reasonable predictions can be made regarding future advances.

Data Abstraction

One of the keys to successful programming is the concept of abstraction. Abstraction is the crucial to building complex software systems. A good definition of abstraction comes from, and can be summed up as concentrating on relevant aspects of the problem and ignoring those that are not currently important.

The psychological notion of abstraction permits one to concentrate on a problem at some level of generalization without regard to irrelevant low-level details; use of abstraction also permits one to work with concepts and terms that are familiar in the problem environment without having to transform them to an unfamiliar structure.

As the size of our problems grows, the need for abstraction dramatically increases. In simple systems, characteristic of languages used in the 1950s and 1960s, a single programmer could understand the entire problem, and therefore manipulate all program and data structures. Programmers today are unable to understand all of the programs and data — it is just too large. Abstraction is required to allow the programmer to grasp necessary concepts. To understand how abstraction works, it is helpful to show the topology or mapping of a language to the data structures and program modules that the language provides. Once we see the topology of early languages, we can better understand the problems and solutions.

1.2.1 Assembly and Machine Languages

The most basic language of a computer, the machine language, provides program instructions in the bits. Even though most computers perform the same kinds of operations, the designers of computer may have chosen different sets of binary codes to perform the operation. Therefore, the machine language of one machine is not necessarily the same as the machine language of another machine. The only consistency among computers is that in any modern computer, all data is stored and manipulated as binary codes.

Assembler is a program that translates a program written in assembly language into an equivalent program in machine language.

Early languages had little distinction between programs and data (see Figure 1-2). The data and program co-existed. Because the data and program often were ill-defined, the boundary was irregular. It often was hard to distinguish between data and code. If a true hacker needed to use the number 62, and he or she knew that the machine instruction was coded as a hex 3E — which is equal to decimal 62 — the hacker would reference the instruction elsewhere in the program to refer to decimal 62.

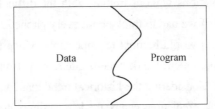

Figure 1-2　The topology of assembly and machines languages

1.2.2　Early Languages

The first programming languages widely used had a clear separation between the data and the program. These languages had a global data structure, but permitted modularization of program structure. Typically, there was only single-level modularization of the program (see Figure 1-3). While this separation of data and program was a good thing, all program segments were at a single level, and typically referenced each other in very complex ways. In software engineering terms, the cohesion was typically very low, and the coupling was quite high. In other words, modules tended to perform many tasks, and there was a lot of dependence on the workings of other modules. In addition, each module had unlimited access to all data because the data was global to all modules. Global data is bad — it makes maintenance extremely difficult, since it is hard to determine which module is 'trashing' the data.

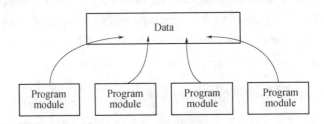

Figure 1-3　The topology of early languages (1950s and 1960s)

To build larger and larger systems, which were more and more complex, improvements were needed to make large-scale development easier.

1.2.3 Later-Generation Languages

Most of us who learned to program in the 1970s and '80s learned to program with languages such as PL/1, Pascal, ALGOL, later versions of FORTRAN, or C. These provided hidden modules within larger modules, which led to an increase in cohesion; it was easier to put logically related routines together and a decrease in coupling. This resulted in a marked increase in maintainability and robustness.

See Figure 1-4 for an example of the topology of these languages. However, the same problems that existed for earlier languages remained, because there were no improvements on the data's topology. Because the data was unsegmented and unprotected — all modules that had access had full access to read and modify the data — problems still existed.

What was needed was a way to enforce limitations on the data so that effective abstractions could be created and enforced.

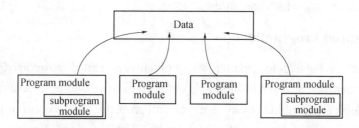

Figure 1-4   The topology of later languages (1970s and 1980s)

1.2.4 Modern Languages

Most of us who learned to program in the 1990s were exposed to languages such as Ada, C++, Java, or C#. Languages such as these permit abstractions in both the data and the program units, which in turn permit us to not only create abstractions, but also enforce the abstractions. In Ada and Ada 95, the concept of private types permit sharing of data that has limited, and compiler-enforced, restrictions on the use of the data. In C++ and Java, classes can define method access, such as public and private (refer to Figure 1-5). These modifications to the way we access data allow us to create powerful abstractions, and then control the way the data is accessed. We can even control who can access the data, and limit access to certain program modules.

What did the increasing complexity of topology gain us? To ask it another way, how has the evolution of program and data abstractions helped us do our job? The answer: it has allowed us to "abstract away" more and more low-level knowledge about the solution domain, and concentrate more on the solution domain.

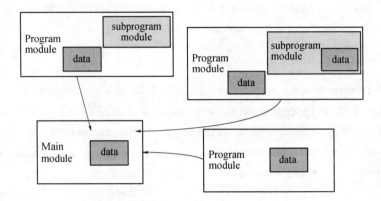

Figure 1-5   The topology of modern languages

1.3 Programming Methodologies

Two popular approaches to programming design are the structured approach and the object-oriented approach, which are outlined below.

1.3.1 Structured Programming

Dividing a problem into smaller sub-problems is called ***structured design***. Each sub-problem is then analysed and a solution is obtained to solve the sub-problem. The solutions to all sub-problems are combined to solve the overall problem. This process of implementing a structured design is called ***structured programming***. The structured-design approach is also known as top-down design, stepwise refinement and modular programming (Figure 1-6).

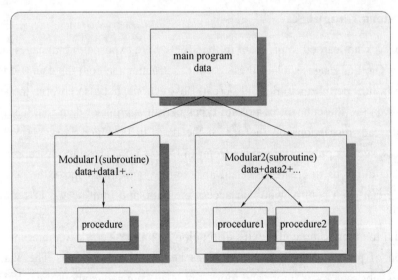

Figure 1-6   Structured programming (or modular programming)

For using structured programming, there are many good reasons: codes are easier to understand errors are easier to find. An error will always be localized to a subroutine or function rather than buried somewhere in a mass of code scoping of variables can be controlled more easily reuse of code—as well as being reused within a single application, modular programming allows code to be used in multiple applications programs are easier to design—the designer just needs to think about high level functions collaborative programming is possible—modular programming enables more than one programmer to work a single application at the same time code can be stored across multiple files. Now, we write a simple program which calculates the circumferences and areas of diffifferent sized circles. We design it in two ways.

Example 1-2: A simple example without structured programming.

```
/* File:example1_2.c
 * The program calculates circumferences and areas of different sized circles without using
 * structured programming.
 */
1      #include <stdio.h>
2      #define PI   3.14156926
3      int main()
4      {
5          float r1 = 1.0
6          float c1 = 2 * PI * r1;
7          float a1 = PI * r1 * r1;
8          printf ("The circumference is %6.2f, the area is %6.2f ", c1, a1);
9
10         float r2 = 2.0
11         float c2 = 2 * PI * r2;
12         float a2 = PI * r2 * r2;
13         printf (" The circumference is %6.2f, the area is %6.2f ", c2, a2);
14
15         float r3 = 3.0
16         float c3 = 2 * PI * r3;
17         float a3 = PI * r3 * r3;
18         printf (" The circumference is %6.2f, the area is %6.2f ", c3, a3);
17
18         float r4 = 4.0
19         float c4 = 2 * PI * r4;
20         float a4 = PI * r4 * r4;
21         printf ("The circumference is %6.2f, the area is %6.2f ", c4, a4);
22
23         return 0;
24     }
```

Here the same functionality is repeated but does work. However, the programmer can work more efficiently by using structured programming.

The aim of the programmer is now to ensure that the code is reused as much as possible. They do this by identifying the code that can be placed separate functions and subroutines.

Example 1-3: A simple example with structured programming.

```
/* File: example1_3.c
 * The program calculates circumferences and areas of different sized circles with using
 * structured programming.
 */
1      #include <stdio.h>
2      #define  PI   3.14156926
3
4      float circum (float r)
5      { return 2 * PI * r;  }
6      float area (float r)
7      { return PI * r * r; }
8      void print (float r)
9      {
10         printf(" The circumference is %6.2f, the area is %6.2f ", circum(r), area(r));
11     }
12
13     int main()
14     {
15         print(1.0);
16         print(2.0);
17         print(3.0);
18         print(4.0);
19         return 0;
20     }
```

The output is the same as before, but this time the chance of the operation of the script being altered due to a typing error is greatly reduced, and if there is an error then the task of correcting it is made much simpler. The programmer can also use the functionality throughout their application without worrying about having to rewrite the code, thereby improving the efficiency of both the code and their time.

1.3.2 Object-Oriented Programming

Object-oriented programming (OOP) is a widely used programming methodology(Figure1-7). In OOP, the first step is to identify the components called ***objects***, which form the basis of the solution, and to determine how these objects interact with one another. The next step is to specify for each object the relevant data and possible operations to be performed on the data.

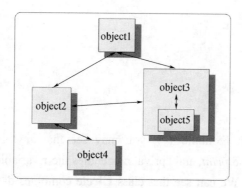

Figure 1-7   Object-oriented programming

Object-oriented programming is centered on the object. An object is a programming element that contains data and the procedures that operate on the data. The objects contain, within themselves, both the information and the ability to manipulate the information.

Let us change the program in Example 1-3 using object-oriented programming.

Example 1-4:   A simple example with OOP.

```
/* File: example1_4.cpp
 * The program calculates circumferences and areas of different sized circles with using OOP.
 */
1     #include <stdio.h>
2     #define   PI   3.14156926
3
4     class Circle{
5     public:
6         Circle(float R)
7         {    r = R; }
8         float circum()
9         { return 2 * PI * r; }
10        float area()
11        { return PI * r * r; }
12        void print()
13        {
14            printf (" The circumference is %6.2f, the area is %6.2f ", circum(), area());
15        }
16    private:
17        float r;
18    };
19
20    int main()
21    {
22        Circle c1(1.0), c2(2.0), c3(3.0), c4(4.0);
```

```
23          c1.print();
24          c2.print();
25          c3.print();
26          c4.print();
27          return 0;
28      }
```

From first glance in Example 1-4, you may see that anything after "public:" is the functions: *circum*, *area* and *print*, and "private:" denotes the **r** variable. This is the definition of an ***object***, called a ***class***. We can see that class **Circle** combines data and operations on data into a single unit. In a main function, we create four different objects of class **Circle**, i.e. *c1*, *c2*, *c3* and *c4*, and deal with them directly.

The primary differences between the two approaches are their use of data. In a structured program, the design centers on the rules or procedures for processing the data. The procedures, implemented as functions in C++, are the focus of the design. The data objects are passed to the functions as parameters. The key question is how the functions will transform the data they receive for either storage or further processing. In an object-oriented program, the design centers on the objects that contain the data and the necessary functions to process the data. Procedural programming has been the mainstay of computer science since its beginning. However, object-oriented programming is used in main stream program method today.

Think it Over
Why uses classes in OOP instead of functions?

1.4 Object-Oriented Programming

Object-oriented programming (OOP) is a programming paradigm that uses "objects" —data structures consisting of data fields and methods together with their interactions—to design applications and computer programs. Programming techniques may include features such as data abstraction, encapsulation, modularity, information, polymorphism, and inheritance. It was not commonly used in mainstream software application development until the early 1990s. Many modern programming languages now support OOP.

Object-oriented programming has roots that can be traced to the 1960s. As hardware and software became increasingly complex, manageability often became a concern. Researchers studied ways to maintain software quality and developed object-oriented programming in part to address common problems by strongly emphasizing discrete, reusable units of programming logic. The technology focuses on data rather than processes, with programs composed of self-sufficient modules ("classes"), each instance of which ("objects") contains all the information needed to manipulate its own data structure ("members"). This is in contrast to the existing modular programming which had been dominant for many years that focused on the *function* of a module, rather than specifically the data, but equally provided for code reuse, and self-sufficient reusable units of programming logic, enabling collaboration through the use of linked modules (subroutines).

An object-oriented program may thus be viewed as a collection of interacting *objects*, as opposed to the conventional model, in which a program is seen as a list of tasks (subroutines) to perform. In OOP, each object is capable of receiving messages, processing data, and sending messages to other objects. Each object can be viewed as an independent "machine" with a distinct role or responsibility.

Class

A Class is a user defined data type which contains the variables, properties and methods in it. A class defines the abstract characteristics of a thing (object), including its characteristics (its **attributes**, **fields** or **properties**) and the thing's behaviors (the things it can do, or **methods**, **operations** or **features**). One might say that a class is a *blueprint* or *factory* that describes the nature of something.

For example, the class Dog would consist of traits shared by all dogs, such as breed and fur color (characteristics), and the ability to bark and sit (behaviors). Classes provide modularity and structure in an object-oriented computer program. A class should typically be recognizable to a non-programmer familiar with the problem domain, meaning that the characteristics of the class should make sense in context. Also, the code for a class should be relatively self-contained (generally using **encapsulation**). Collectively, the properties and methods defined by a class are called **members**.

Method

Method is a set of procedural statements for achieving the desired result. It performs different kinds of operations on different data types. In a programming language, methods (sometimes referred to as "functions") are verbs. For example, Lassie, being a Dog, has the ability to bark. So *bark* is one of Lassie's methods. She may have other methods as well, for example *sit* or *eat* or *walk* or save (Timmy). Within the program, using a method usually affects only one particular object; all Dogs can bark, but you need only one particular dog to do the barking.

Message passing

Message passing is the process by which an object sends data to another object or asks the other object to invoke a method which is known to some programming languages as interfacing. For example, the object called Breeder may tell the Lassie object to sit by passing a "sit" message which invokes Lassie's "sit" method.

Abstraction

Abstraction is one of the most powerful and vital features provided by object-oriented programming. The concept of abstraction relates to the idea of hiding data that are not needed for presentation. The main idea behind data abstraction is to give a clear separation between properties of data type and the associated implementation details. This separation is achieved in order that the properties of the abstract data type are visible to the user interface and the implementation details are hidden. Thus, abstraction forms the basic platform for the creation of user-defined data types called objects. Data abstraction is the process of refining data to its essential form. An Abstract Data Type (ADT) is defined as a data type that is defined in terms of the operations that it supports and not in terms of its structure or implementation.

In object-oriented programming language C++, it is possible to create and provide an interface that accesses only certain elements of data types. The programmer can decide which user to give or grant access to and hide the other details. This concept is called data hiding which is similar in concept to data abstraction.

For example, in a touch screen at railway station or ATM machine in the bank, we just use the touch screen application to satisfy our needs, we don't see what is happening inside its software or about its OS.

Encapsulation

Encapsulation conceals the functional details of a class from objects that send messages to it.

For example, the Dog class has a *bark* method. The code for the *bark* method defines exactly how a bark happens (e.g., by *inhale* method and then an *exhale* method, at a particular pitch and volume). Timmy, Lassie's friend, however, does not need to know exactly how she barks. Encapsulation is achieved by specifying which classes may use the members of an object. The result is that each object exposes to any class a certain *interface* — those members accessible to that class. The reason for encapsulation is to prevent clients of an interface from depending on those parts of the implementation that are likely to change in the future, thereby allowing those changes to be made more easily, that is, without changes to clients. For example, an interface can ensure that puppies can only be added to an object of the class Dog by code in that class. Members are often specified as **public**, **protected** or **private**, determining whether they are available to all classes, sub-classes or only the defining class.

Inheritance

Inheritance is a process in which a class inherits all the state and behavior of another class. This type of relationship is called child-Parent or is-a relationship. "Subclasses" are more specialized versions of a class, which *inherit* attributes and behaviors from their parent classes, and can introduce their own.

For example, the class Dog might have sub-classes called Collie, Chihuahua, and GoldenRetriever. In this case, Lassie would be an instance of the Collie subclass. Suppose the Dog class defines a method called *bark* and a property called furColor. Each of its sub-classes (Collie, Chihuahua, and GoldenRetriever) will inherit these members, meaning that the programmer only needs to write the code for them once.

Each subclass can alter its inherited traits. For example, the Collie subclass might specify that the default furColor for a collie is brown-and-white. The Chihuahua subclass might specify that the *bark* method produces a high pitch by default. Subclasses can also add new members.

The Chihuahua subclass could add a method called *tremble*. So an individual chihuahua instance would use a high-pitched *bark* from the Chihuahua subclass, which in turn inherited the usual *bark* from Dog. The chihuahua object would also have the *tremble* method, but Lassie would not, because she is a Collie, not a Chihuahua. In fact, inheritance is an "*a*...**is a**" relationship between classes, while instantiation is an "**is a**" relationship between an object and a class: *a* Collie *is a* Dog ("a... is a"), but Lassie *is a* Collie ("is a"). Thus, the object named Lassie has the methods from both classes Collie and Dog.

Polymorphism

Polymorphism allows the programmer to treat derived class members just like their parent class's members. More precisely, Polymorphism in object-oriented programming is the ability of objects belonging to different data types to respond to calls of methods of the same name, each one according to an appropriate type-specific behavior. One method, or an operator such as +, -, or *, can be abstractly applied in many different situations. If a Dog is commanded to a *speak* method, this may elicit a *bark* method. However, if a Pig is commanded to *speak*, this may elicit an *oink* method. Each subclass overrides the *speak* method inherited from the parent class Animal.

1.5 C++ Programming Language

1.5.1 History of C and C++

C evolved from two previous programming language BCPL and B. BCPL was developed in 1967 by Martin Richards as a language for writing operating systems software and compilers. Ken Thompson modeled many features in his language B after their counterparts in BCPL and used B to create early versions of the UNIX operating system at BELL Laboratories in 1970 on a DEC PDP-7 computer. Both BCPL and B were "typeless" language—every data item occupied one "word" in memory and burden of treating a data item as a whole number or a real number, for example, fell on the shoulders of programmer.

The C language was evolved from B by Dennis Ritchie at Bell Laboratories and was originally implemented on a DEC PDP_11 computer in 1972. C uses many important concepts of BCPL and B while adding data typing and other features. C initially became widely known as the development language of the UNIX operating system. Today, most operating systems are written in C and/or C++.

By the late 1970s, C has evolved into what is now referred to as "traditional C","classic C", or "Kernighan and Ritchie C". The widespread use of C with various types of computers (sometimes called hardware platforms) unfortunately led to many various. There were similar, but often incompatible. This was a serious problem for program developers who needed to write portable programs that would run on several platforms. It became clear that a standard version of C was needed. In 1983, the X3J11 technical committee was created under the American Nation Standard Committee on Computers and Information Processing (X3) to "provide an unambiguous and machine-independent definition of the language". In 1989, the standard was approved. ANSI cooperated with the International Standards Organization (ISO) to standardize C worldwide, the joint standard document was published in 1990 and is referred to as ANSI/ISO 9899:1990. Copies of this document may be ordered from ANSI. The second edition of Kernighan ad Ritchie, published in 1988, reflects this version called ANSI C, a version of the language now used worldwide (Ke88).

C++, an extension of C, was developed by Bjarne Stroustrup at Bell Laboratories during 1983-1985. Prior to 1983, Bjarne Stroustrup added features to C and formed what he called "C with classes". He had combined the Simula's use of classes and object-oriented features with the power and efficiency of C. The term C++ was first used in 1983.

The name C++ was coined by Rick Mascitti in the summer of 1983. The name signifies the evolutionary nature of the changes from C; "++" is the C increment operator. The slightly shorter name "C+" is a syntax error; it has also been used as the name of an unrelated language. Connoisseurs of C semantics find C++ inferior to ++C. The language is not called D, because it is an extension of C, and it does not attempt to remedy problems by removing features.

Like C, ISO also published the first international standard for C++ in 1998, known as C++98. C++98 includes the Standard Template Library (STL) which began its conceptual development in 1979. In 2003 and 2005, ISO respectively revised problems in C++98. A new C++ stardard (known as C++11) was approved by ISO in 2011. C++ 11 added new features into the core language and standard library. These new features are very useful for advance C++ programming.

1.5.2 Learning C++

As Bjarne Stroustrup mentioned, C++ is a general-purpose programming language with a bias towards systems programming that
- is a better C,
- supports data abstraction,
- supports object-oriented programming, and
- supports generic programming.

The most important thing to do when learning C++ is to focus on concepts and not get lost in language-technical details. The purpose of learning a programming language is to become a better programmer, that is, to become more effective at designing and implementing new systems and at maintaining old ones. For this, an appreciation of programming and design techniques is far more important than an understanding of details; that understanding comes with time and practice.

C++ supports a variety of programming styles. All are based on strong static type checking, and most aim at achieving a high level of abstraction and a direct representation of the programmer's ideas. Each style can achieve its aims effectively while maintaining run-time and space efficiency. A programmer coming from a different language (say C, Fortran,

Smalltalk, Pascal, or Modula-2) should realize that to gain the benefits of C++, they must spend time learning and internalizing programming styles and techniques suitable to C++. The same applies to programmers used to an earlier and less expressive version of C++.

Thoughtlessly applying techniques effective in one language to another typically leads to awkward, poorly performing, and hard-to-maintain code. Such code is also most frustrating to write because every line of code and every compiler error message reminds the programmer that the language used differs from "the old language." You can write in the style of Fortran, C, Smalltalk, etc., in any language, but doing so is neither pleasant nor economical in a language with a different philosophy. Every language can be a fertile source of ideas of how to write C++ programs. However, ideas must be transformed into something that fits with the general structure and type system of C++ in order to be effective in the different context.

C++ supports a gradual approach to learning. How you approach learning a new programming language depends on what you already know and what you aim to learn. There is no one approach that suits everyone. My assumption is that you are learning C++ to become a better programmer and designer. That is, I assume that your purpose in learning C++ is not simply to learn a new syntax for doing things the way you used to, but to learn new and better ways of building systems. This has to be done gradually because acquiring any significant new skill takes time and requires practice. Consider how long it would take to learn a new natural language well or to learn to play a new musical instrument well. Becoming a better system designer is easier and faster, but not as much easier and faster as most people would like it to be.

Word Tips

abstract	*adj.* 抽象的		crux	*n.* 难点，关键
abstraction	*n.* 抽象，抽取		deem	*vi./vt.* 认为，相信
accurate	*adj.* 精确的		detail	*n.* 细节，小事
actually	*adv.* 实际上		discrete	*adj.* 分离的，不相关的
adage	*n.* 言语，格言		distinct	*adj.* 清晰的，明显的
algorithm	*n.* 运算法则		domain	*n.* 范围，领域
analogy	*n.* 类似，相似，类推		dome	*n.* 圆屋顶
analysis	*n.* 分析，分析报告		dominant	*adj.* 占优势的，突出的
broad	*adj.* 宽的，广的		efficient	*adj.* 有能力的，效率高的
calculate	*vi./vt.* 计算，估计		emphasis	*n.* 强调，重点
chihuahua	*n.* 吉娃娃狗		encapsulation	*n.* 封装
clarify	*v.* 使清楚		evolution	*n.* 演变，进化，发展
client	*n.* 委托人，顾客		gigantic	*adj.* 巨大的，庞大的
code	*n.* 代码		implement	*vt.* 执行，实现
cohesion	*n.* 聚合		individual	*adj.* 个别的，单独的
concrete	*adj.* 实体的，有形的		infancy	*n.* 早期
coupling	*n.* 耦合		instruction	*n.* 命令，指示

integer	*n.* 整数	property	*n.* 特性，属性
internal	*adj.* 内部的	psychology	*n.* 心理，心理学
internalize	*v.* 吸收同化	puppy	*n.* 小狗
invoke	*vt.* 调用	pyrrhic	*n.* 出征舞，抑抑格
iteration	*n.* 反复，迭代	refinement	*n.* 精化
logical	*adj.* 逻辑上的	robust	*adj.* 强壮的，健全的
mainstay	*n.* 支柱，骨干	script	*n.* 脚本
maintain	*vt.* 维护	segment	*n.* 部分，片段
methodology	*n.* 方法	separation	*n.* 分离，分开
modular	*adj.* 模块的	sequence	*n.* 顺序
module	*n.* 单元，单位	specification	*n.* 说明，详述
multiple	*adj.* 多重的，多种多样的	specify	*vt.* 详述
paradigm	*n.* 范型	syntax	*n.* 句法，句法规则
parameter	*n.* 参数	temporary	*adj.* 临时的，暂时的
phase	*n.* 阶段，时期	thereby	*adv.* 由此，因而
phonetic	*adj.* 语言的	thoughtlessly	*adv.* 轻率地，草率地
platform	*n.* 平台	topology	*n.* 拓扑结构
polymorphism	*n.* 多态性	trace	*vt.* 追踪，发现，找到
prediction	*n.* 预言	tremble	*vi.* 发抖，颤动
premise	*n.* 前提	unambiguous	*adj.* 不含糊的，清楚
procedure	*n.* 程序，过程		

Exercises

1. What is planning the performance of a task or an event?
2. What is a step-by-step procedure for solving a problem in a finite amount of time?
3. What is OOP? What is the name of the data strueture we used? What are the components of this data structure?
4. What are the characteristics of OOP?
5. Explain the difference between the structured programming and object-oriented programming.
6. Why do we need object-oriented programming?

Chapter 2
Basic Facilities
— *Shifting From C to C++ Programs* —

Let's all move one place on.
—Lewis Carroll

Objectives
- To be able to construct a simple C++ program
- To be able to create and recognize legal C++ identifiers
- To understand how to construct programs modularly from small pieces called functions
- To understand the mechanism used to pass information between functions
- To be able to use references to pass arguments to functions

2.1　C++ Program Structure

The C++ program consists of indertifiers, declarations, variables, constants, expressions, statements, and comments. A common structure for simple, one file, C++ program includes as follows:

Comments describe program feature, author, ...
Include statements specify the header files for libraries.
Using namespace statement. Typically, this is only *using namespace std;*.
Global declarations (consts, types, variables, ...). Avoid global variables if possible.
Function prototypes (declarations).
Main function definition.
Function definitions.

The following example is used to illustrate the structure of a simple C++ program. Do not be too concerned with the details in the program—just observe its over-all look and structure.
Example 2-1: A C++ program.

```
//---------------------------------------------------------------
// File: example2_1.cpp
// This program describes the structure of a simple C++ program.
// The program calculates the square of an integer.
// Jiamin Liu   2013-07-10
```

```
//--------------------------------------------------------------------
1    //==includes==
2    #include <iostream>
3    using namespace std;
4    //==function prototype ==
5    int Square ( int);
6    // ==main function==
7    int main()
8    {
9        int n;
10       cout << "Enter an integer to see its square. " << endl;
11       cin >> n;
12       cout << " The square of "  << n << " is " << Square(n) << endl;
13       return 0;
14   }
15   //= functions =
16   int Square(int m)
17   {
18       return m * m;
19   }
```
Result:
```
1    Enter an integer to see its square
2    The square of 4 is 16
```

Comments are used for better understanding of the program statements. The comment entries start with a // symbol on a line and terminate at the end of the line, *called a line comment*. In Example 2-1, the statements

> // File: example2_1.cpp
> // This program describes the structure of a simple C++ program.
> // The program calculates the square of an integer.
> // Jiamin Liu   2013-07-10

Like the C language, the multiple line comment entries can be included in a C++ program using the /* and the */ symbols.

The **#include** directive instructs the compiler to include the contents of the file enclosed within angular brackets into the source file. In Example 2-1, the file *iostream* is included in the source file *example2_1.cpp*.

All C ++ programs comprise one or more functions, which are a logical grouping of one or more statements. A *function* is identified by a *function* name and *function* body.

A function name is identified by a word followed by parentheses. In Example 2-1, *main* is a function name. All programs must have a function called *main*. The execution of a program begins with the *main* function. The keyword *int* along with the function name signifies that the function returns an integer value.

A function body is surrounded by curly braces ({}). The braces delimit a block of program statements. Every function must have a pair of braces.

Input / Output Statements

The statements

    cin >> n;

    cout << "The square of " << n << " is " << Square(n) << endl;

enter a value from keyboard to variable n and cause the enclosed text to be displayed on the screen.

2.2 Input / Output Streams

A program performs three basic operations: it gets data, it manipulates the data, and it output the results. Because writing programs for I/O is quite complex, C++ offers extensive support for I/O operations. In C++, I/O is sequence of bytes, called a *stream*, from the source to the destination.

> A **stream** is a sequence of characters from the source to the destination. There are two types of streams:
> **Input stream** is a sequence of characters from an input device (e.g. keyboard) to the computer.
> **Output stream** is a sequence of characters from the computer to an output device (e.g. monitor).

Input Stream

The standard library offers **istream** for input. The **istream** deal with character string representations of built-in types and can easily be extended to cope with user-defined types.

With input streams, the **extraction operator (>>)** is used to remove values from the stream. This makes sense: when the user presses a key on the keyboard, the key code is placed in an input stream. Your program then extracts the value from the stream so it can be used.

The extraction operator >> ("get from") is used as an input operator; **cin** (read see-in) is one of the standard Input/Output streams, ie. iostream. The type of the right-hand operand of >> determines what input is accepted and what is the target of the input operation.

The following statements demonstrate the input stream using *cin*.

```
1   void f( )
2   {
3       int i;
4       cin >> i;    //read an integer into variable i
5       double d;
6       cin >> d;    //read a double-precision, floating-point number into variable d
7   }
```

The statements in Lines 4 and 6 read an integer and floating-point number, such as 1234 and 34.6, from the standard input into the integer variable **i** and the double-precision, floating-point variable **d**. You find out that *cin* can read the data directly into any variable type.

Object-Oriented Programming in C++

Using *cin* makes the programmer maniplute data easily and flexibly without the consideration of the variable type.

Output Stream

With output streams, the **insertion operator (<<)** is used to put values in the stream. This also makes sense: you insert your values into the stream, and the data consumer (e.g. monitor) uses them.

The insertion operator (<<) ("put in") are used as an output operation. **cout** (read see-out) is also one of the standard Input/Output streams, ie. iostream. By default, values output to **cout** are converted to a sequence of characters.

The following statements are examples of using **cout**:

```
1   void f( )
2   {
3       double d =5.6;
4       cout  << 10 << endl;
5       cout  << d  << endl;
6       cout  << "Hello World! " << endl;
7   }
```

Result:

```
1   10
2   5.6
3   Hello World!
```

Think It Over
What are the differences between the I/O statements of *cin* / *cout* and *scanf* / *printf*?

2.3 Constant

C++ offers the concept of a user-defined constant, a ***const***, to express the notion that the value of a variable cannot be changed directly. This is useful in several contexts. For example, many variables do not actually have their values changed after initialization, symbolic constants lead to more maintainable code than do literals embedded directly in code, pointers are often read through but never written through, and most function parameters are read but not written to.

The keyword ***const*** specifies that a variable's value is constant and tells the compiler to prevent the programmer from modifying it. It can be added to the declaration of a variable to make the variable declared a constant.

For example,

```
1   const  int   MaxN = 90;              // MaxN is a const
2   const  int   v[] = {1, 2, 3, 4};     //v[i] is a const
3   const  int   x;                      //error: no initialization
```

```
4    void f( )
5    {
6        MaxN =200;              //error
7        v[2]++;                 //error
8    }
```

 Because a constant cannot be assigned to, it must be initialized. Declaring variables as *const* ensures that its value will not change within its scope.

Difference Between the *define* Directive and the *const* Statement

In C++, you can use the **const** keyword instead of the *#define* pre-processor directive to define constant values. Values defined with **const** are subject to type checking, and can be used in place of constant expressions. In C++, you can specify the size of an array with a **const** variable as follows:

```
const int MaxN = 100;
char strV[MaxN];                // allowed in C++; not allowed in C
```

In C, constant values default to external linkage, so they can appear only in source files. In C++, constant values default to internal linkage, which allows them to appear in header files.

Depending on how smart it is, a compiler can take advantage of an object being a constant in several ways. For example, the initializer for a constant is often (but not always) a constant expression; if it is, it can be evaluated at compile time. Further, if the compiler knows every use of the const, it need not allocate space to hold it.

For example,

```
const  int  c1 = 1;
const  int  c2 = 2;
const  int  c3 = my_f(3);       //don't know the value of c3 at compile time
extern const int c4;            //don't know the value of c4 at compile time
const  int *p = &c2;            //need to allocate space for c2
```

Given this, the compiler knows the values of **c1** and **c2** so that they can be used in constant expressions. Since the values of **c3** and **c4** are not known at compile time, storage must be allocated for **c3** and **c4**. Because the address of **c2** is taken (and presumably used somewhere), storage must be allocated for c2. The simple and common case is the one in which the value of the constant is known at compile time and no storage needs to be allocated; **c1** is an example of that. The keyword extern indicates that **c4** is defined elsewhere.

Pointers and Constants

When using a pointer, two variables are involved: the pointer itself and the variable pointed to. "Prefixing" a declaration of a pointer with **const** makes the variable, but not the pointer, a constant. To declare a pointer itself, rather than the object pointed to, to be a constant, we use the declaratory operator **const* instead of plain*.

Consider the following statements:

```
1   void f(char* p)
2   {
3       char s[] = "Gorm";
4       const char* pc = s;              //pointer to constant
5       pc[3] = 'g';                     //error: pc points to constant
6       pc = p;                          //ok
7
8       char* const cp = s;              //constant pointer
9       cp[3] = 'a';                     //ok
10      cp = p;                          //error
11
12      const char* const cpc = s;       //const pointer to const
13      cpc[3] = 'a';                    //error
14      cpc = p;                         //error
15  }
```

Note that the statements in Lines 4, 8 and 12 produce different results. Variable **pc** in Line 4 is a pointer to a constant. This means you cannot change the value of variable **pc** whose address it is holding. **cp** in Line 8 is a constant pointer. This means that you cannot change the address pointer **cp** is pointing to, but you can change its value. **cpc** in Line 12 is a constant pointer to a constant.   this means that you cannot change both the address **cpc** is pointing to and the value whose address **cpc** is holding.

2.4 Functions

Let us look back the program in Example 1-3 of Chapter 1. The program is divided into a few blocks, called *function body*(when you solve a program). Sometimes, it is not practical put the entire programming into one function, e.g. a *main* function, as you see in Example 1-3. You must learn to break the problem into manageable pieces.

Functions are like building blocks. This lets you divide complicated problems into manageable pieces. The aims for using functions in a program are to:
- be more brief
- maintain easily
- enhance the efficiency of programming
- be reusable

2.4.1 Function Declarations

A function *declaration* establishes the name of the function, the type of data returned by the function and the number and types of its parameters. The syntax of a function declaration is:

returnValueType functionName (formal parameter list);

A function declaration consists of a return type, a name, and a parameter list. In addition, a function declaration may optionally specify the function's linkage. In C++, the declaration can also specify an exception specification, a const-qualification, or a volatile-qualification.

A declaration informs the compiler of the format and existence of a function prior to its use. A function can be declared several times in a program, provided that all the declarations agree. Implicit declaration of functions is not allowed: every function must be explicitly declared before it can be called.

For example,

```
1   double sqrt(double);           // declaration of the sqrt function with a parameter of double type
2   double  sr2 = sqrt(2);         // invoke the sqrt function by passing the argument of double(2)
3   double  sq3 = sqrt("three");   // error: sqrt() require an argument of double
```

A *function argument* (i.e. *actual parameter*) is an expression or a variable that you use within the parentheses of a function call. A *function parameter* (formal parameter) is an object or reference declared within the parenthesis of a function declaration or definition. When you call a function, the arguments are evaluated, and each parameter is initialized with the value of the corresponding argument. The semantics of argument passing are identical to those of assignment.

Some declarations do not name the parameters within the parameter lists; the declarations simply specify the types of parameters and the return values. This is also called *prototype*. A function prototype consists of the function return type, the name of the function, and the parameter list. For example, the following statement describes this:

>   double sqrt(double);

Function prototypes are required for compatibility between C and C++. The non-prototype form of a function that has an empty parameter list means that the function takes an unknown number of parameters in C, whereas in C++, it means that it takes no parameters.

2.4.2 Function Definitions

Every function in a program must be defined somewhere (**once only**) before it is invoked. A function definition is a function declaration in which the body of the function is presented. The syntax of a function definition is:

>   returnValueType   functionName (parameter list)
>   {   statement (i.e. function body)   }

Example 2-2: The declaration and definition of function *swap*.

//---
// File: **example2_2.cpp**
// This program defines a *swap* function which exchanges two integers.
//---

```
1   #include <iostream>
2   using namespace std;
3
```

```
4   void  swap (int *,  int*);          // function declaration
5
6   void  swap (int *p,  int *q)        // function definition
7   {
8       int  t = *p;
9       *p = *q;
10      *q = t;
11  }
12
13  int main()
14  {
15      int a = 3, b = 5;
16      cout << "a = " << a << " b = " << b << endl;
17      swap(a,b);
18      cout << "a = " << a << " b = " << b << endl;
19      return 0;
20  }
```

The types of definitions and declarations for a function must specify the same type. The parameter names, however, are not part of the type and need not be identical. It is not uncommon to have function definitions with unused parameters.

```
void   search (table *t, const char * key, const char*)
{
        //no use of the third parameter
}
```

As shown, the fact that a parameter is unused can be indicated by not naming it. Typically, unnamed parameters arise from the simplification of code or from planning ahead for extensions. In both cases, leaving the parameter in place, although unused, ensures that callers are not affected by the change.

2.4.3 Default Arguments

A general function often needs more parameters that are necessary to handle simple cases. In particular, functions that construct objects often provide several options for flexibility.

A **default parameter** is a function parameter that has a default value provided to it. If the user does not provide a value for this parameter, the default value will be used. Of course, the user can change the default value if he or she provides a value for the default parameter.

Consider a function for printing an integer in Example 2-3. Giving the user an option of what **value2** to print it in seems reasonable, but in most programs integers will be printed as decimal integer values.

Example 2-3: A print function using default arguments.

//---
// File: **example2_3.cpp**

// This program defines a print function with default arguments.
// ---
```
1   #include <iostream>
2   using namespace std;
3
4   void print (int value1, int value2 = 10);      // default vaule of parameter value2 is 10
5
6   int main ()
7   {
8       print(31);
9       print(32, 30);
10      return 0;
11  }
12
13  void print(int value1, int value2)
14  {
15      cout<< value1 << "   " << value2 <<endl;
16  }
```
Result:
```
1   31   10
2   32   30
```

In the first function call in Line 8, the caller does not supply an argument for variable **value2**, so the function uses the default value of 10. In the second call in Line 9, the caller does supply a value for variable **value2**, so the user-supplied value is used.

A default argument is type checked at the time of the function declaration and evaluated at the time of the call. Default arguments may be provided for trailing parameters only. For example, here is a function prototype for which multiple default argument might be commonly used.

Example 2-4: A printing function with multiple default arguments.
//---
// File: **example2_4.cpp**
// This program defines a print function with multiple default arguments.
//---
```
1   #include <iostream>
2   using namespace std;
3
4   void print (int value1 = 10, int value2 = 20, int value3 = 30);
5   int  main ()
6   {
```

```
7              print();
8              print(30);
9              print(30, 40);
10             print(30, 40, 50)
11
12     }
13     void print(int value1, int value2, int value3)
14     {
15         cout << value1 << "   " << value2 << "   " <<value3<<endl;
16     }
```
Result:
```
1    10  20  30
2    30  20  30
3    30  40  30
4    30  40  50
```
In Example 2-4, you maybe notice the statements in Lines 4 and 13. They indicate the function declaration with default values in Line 4 and function definition without default values.

It is impossible to supply a user-defined value for **value3** without also supplying a value for **value1** and **value2**. This is because C++ does not support a function call, such as, *print*(, , 3).

This has two major consequences:

- All default parameters within declaration or definition must be the rightmost parameters. The following is not allowed:

    void   print(int value1 = 0, int value2= 0, int value3);   //error
    void   print(int value1 = 0, int value2, int value3 = 0);   // error
    void   print(int value1,   int   value2 = 0, int value3 = 0);   //ok

- The leftmost default parameter should be the one most likely to be changed by the user when the function is called.

2.4.4   Inline Functions

An inline function is one for which the compiler copies the code from the function definition directly into the code of the calling function rather than creating a separate set of instructions in memory. Instead of transferring control to and from the function code segment, a modified copy of the function body may be substituted directly for the function call. In this way, the performance overhead of a function call is avoided.

A function is declared inline by using the **inline** function specifier or by defining a member function within a class or structure definition. The **inline** specifier is only a suggestion to the compiler that an inline expansion can be performed; the compiler is free to ignore the suggestion.

The following statements show an inline function definition.

    inline int maxValue (int   n, int m)

    {

        return (n < m) ? m : n;

    }

The **inline** specifier is a hint to the compiler that it should attempt to generate code for a call of function *maxValue* inline rather than laying down the code for the function once and then calling through the usual function call mechanism.

The use of the **inline** specifier does not change the meaning of the function. However, the inline expansion of a function may not preserve the order of evaluation of the actual arguments. Inline expansion also does not change the linkage of a function: the linkage is external by default.

2.4.5 Overloading Functions

Most often, it is a good idea to give different functions different names, but when some functions conceptually perform the same task on objects of different types, it can be more convenient to give them the same name. Using the same name for operations on different types is called ***overloading***.

The technique is already used for the basic operations in C++. That is, there is only one name for addition, +, yet it can be used to add values of integer, floating-point, and pointer types. This idea is easily extended to functions defined by the programmer. Several functions have the same name. This is called ***overloading a function name*** or ***function overloading***.

Example 2-5: Overloading functions.

//---
// File: **example2_5.cpp**
// This program defines three print functions with the same name, each performing
// a similar operation on a different data type.
//---

```
1     #include <iostream>
2     using namespace std;
3
4       void print(int i)
5       {
6            cout << " Here is int " << i << endl;
7       }
8       void print(double f)
9       {
10           cout << " Here is double " << f << endl;
11      }
```

```
12
13   void print(char* c)
14   {
15          cout << "Here is char* " << c << endl;
16   }
17
18   int main()
19   {
20          print(10);
21          print(10.10);
22          print("ten");
23          return 0;
24   }
```

Result:

```
1   Here is int 10
2   Here is double 10.10
3   Here is char* ten
```

These *print* functions in Example 2-5 have different parameter lists. Function *print* in Line 4 takes one parameter of type int. Function *print* in Line 8 takes one parameter of type double. Function *print* in Line 13 takes one parameter of type char*. The parameter types of these functions are different. They all overload function name *print*.

 To overload a function name, any two definitions of functions must have different formal parameter lists:
- Different numbers of formal parameters, or
- If the number of formal parameter is the same, then data type of the formal parameters, in the order you list them, must differ in at least one position.

How to Match Function Calls with Overloaded Functions

Given that the following overloading functions are defined.

    void   print (int);
    void   print(const   char*);
    void   print(double);
    void   print(long);
    void   print(char);

An example is given by using the above definitions.

```
//----------------------------------------------------------------
// This program calls overloaded functions with different variable types.
//----------------------------------------------------------------
1   void   main()
2   {
```

```
3      char c='a';
4      int i = 5;
5      short s =7;
6      float f =5.67;
7
8      print(c);              //exact   match:  invoke print (char)
9      print(i);              //exact   match:  invoke print (int)
10     print(s);              //integral promotion: invoke print(int)
11     print (f);             //float to double promotion: print(double)
12     print('a');            //exact match: invoke print(char)
13     print(49);             //exact match: invoke print(int)
14     print(0);              //exact   match: invoke print(int)
15     print("a");            //exact   match: invoke print(const   char*)
16  }
```

The call *print* (0) invokes *print (int)* because 0 is an *int*. The call *print ('a')* invokes *print (char)* because 'a' is a *char*. The reason to distinguish between conversions and promotions is that we want to prefer safe promotions, such as *char* to *int*, over unsafe conversions, such as *int* to *char*.

Therefore, making a call to an overloaded function results in one of three possible outcomes:

(1) A match is found. The call is resolved to a particular overloaded function.

(2) No match is found. The arguments cannot be matched to any overloaded function.

(3) An ambiguous match is found. The arguments matched more than one overloaded function.

When an overloaded function is called, C++ goes through the following process to determine which version of the function will be called:

(1) C++ tries to find an exact match. This is the case where the actual parameter exactly matches the parameter type of one of the overloaded functions. For example, in the above example,

> print(c);
> print(i);
> print(s);
> print(0);

Although integer 0 could technically match *print (char*)*, it exactly matches *print(int)*. Thus *print(int)* is the best match available.

(2) If no exact match is found, C++ tries to find a match through promotion; that is, certain types can be automatically promoted via internal type conversion to other types. The conversions between different types are

- *Char, unsigned char,* and *short* is promoted to an *int*.
- *Unsigned short* can be promoted to *int* or unsigned *int*, depending on the size of an *int*

- *Float* is promoted to *double*
- *Enum* is promoted to *int*

For example,

```
1  void print(char* );
2  void print(int);
3
4  print('a')    //promoted to match print(int)
```

In this case, because there is no *print*(*char*), the char 'a' is promoted to an integer, which then matches *print*(*int*).

(3) If no promotion is found, C++ tries to find a match through standard conversion. Standard conversions include:

- Any numeric type will match any other numeric type, including unsigned (e.g. *int* to *float*)
- Enum will match the formal type of a numeric type (e.g. enum to float)
- Zero will match a pointer type and numeric type (e.g. 0 to char*, or 0 to float)
- A pointer will match a void pointer

For example,

```
1  void print(struct );
2  void print(float);
3
4  print('a')    //promoted to match print(float)
```

In this case, because there is no print(char), and no print(int), the 'a' is converted to a float and matched with print(float).

Ambiguous Matches

If every overloaded function has to have unique parameters, how is it possible that a call could result in more than one match? Because all standard conversions are considered equal, and all user-defined conversions are considered equal, if a function call matches multiple candidates via standard conversion or user-defined conversion, an ambiguous match will result.

For example,

```
1  void print(unsigned int );
2  void print(float);
3
4  print('a')    //error: ambiguous match
```

In the case of *print*('a'), C++ cannot find an exact match. It tries promoting 'a' to an *int*, but there is no *print*(int) either. Using a standard conversion, it can convert 'a' to both an *unsigned int* and a floating point value. Because all standard conversions are considered equal, this is an ambiguous match.

Default Arguments and Overloading Functions

Sometimes, functions with default parameters may be overloaded.

For example, the following is allowed:

```
1    void print(char *sPtr );
2    void print(char c = ' ');
3    print()        //ok: default parameter
4    print(' ');    //ok: match print(char c=' ')
```

In this case, the statement in Line 3 is executed by calling the print function in Line 2. The statement in Line 4 is executed by calling the print function in Line 1.

However, it is important to note that default parameters do NOT count towards the parameters that make the function unique. Consequently, the following is not allowed:

```
1    void print(int value );
2    void print(int value1, int value2 = 10);
3
4    print(10);    //error: ambiguous
```

If the caller were to call *print*(10), the compiler would not be able to disambiguate whether the user wanted *print*(int) or *print*(int, 20) with the default value.

Think These Over

1. What is the difference between a function declaration and a function definition?
2. What is the purpose of using default values in the formal parameter list of the function?
3. If the functions have the same name and formal parameter list, but they have different return types, do you think the function name can be overloaded?

2.5 References

2.5.1 Reference Definition

A **reference** is a simple reference data type. A reference is an alias or alternative name for an object. This means that C++ allows you to create the second name for the object that you can use to read or modify the original data stored in the object. The main use of references is for specifying arguments and return values for functions in general and for overloaded operators in particular.

The syntax of declaring a reference variable is

variableType&   variableName

where **variableType** is a data type and **variableName** is an identifier whose type is reference to variableType.

For example,

```
int    i = 1;
int    &r = i;              // variables r and i now refer to the same int
int    x = r;               // x=1
r = 2;                      // i=2
```

             int &r1;   // *error*

Notice that the statement *int&r = i* means that variable r is a reference type. When variable **r** is declared, it will become an alias for integer variable **i**, that is, the address of variable **r** is the same as that of variable **i**. Whenever you use variable **r**, you can just treat it as though it were a regular integer variable. But when you create it, you must initialize it with variable **i**.

Example 2-6: Definition of reference variable **r**.

//--
// File: **example2_6.cpp**
// This program defines the reference variable r within the main function.
//--

```
1    #include <iostream>
2    using namespace std;
3
4    int main()
5    {
6         int iOne;
7         int &r = iOne;
8         iOne = 5;
9
10        cout << "iOne: " << iOne <<endl;
11        cout << "r: " << r <<endl;
12        cout << "&iOne: " << &iOne <<endl;
13        cout << "&r: " << &r <<endl;
14
15        int iTwo = 8;
16        r = iTwo;
17
18        cout << "iOne: " << iOne <<endl;
19        cout << "iTwo: " << iTwo <<endl;
20        cout << "r: " << r <<endl;
21
22        cout << "&iOne: " << &iOne <<endl;
23        cout << "&iTwo: " << &iTwo <<endl;
24        cout << "&r: " << &r <<endl;
25        return 0;
26   }
```

Result:

1   iOne: 5
2   r: 5
3   &iOne: 0065FDF4

Chapter 2   Basic Facilities

4   &r: 0065FDF4
5   iOne: 8
6   iTwo: 8
7   r: 8
8   &iOne: 0065FDF4
9   &iTwo: 0065FDEC
10   &r: 0065FDF4

In this example, let us assume that memory location 0065FDF4 is allocated for **iOne**, and memory location 0065FDEC is allocated for **iTwo**, as shown in Figure 2-1.

Here, **r** is of type "reference to *int*". The **r** variable refers to an integer variable **iOne**. This means that both variables **r** and **iOne** have the same address in the memory. The **r** variable is changed as changing **iOne**. The statement in Line 8 assigns **r** by variable **iTwo**. We find that the value of **iOne** is changed (see Line 5 of Result).

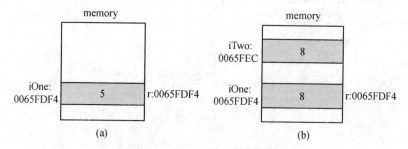

Figure 2-1   Variable **r** in Example 2-6
(a) Definition of variable **r**;   (b) After statement r= iTwo.

References and Pointers

We can find out that a reference variable is used easily like a regular variable. Notice that the reference variables do not need to be allocated a new address in the memory. From this point, we have to consider a pointer variable. The following example illustrates the difference between references and pointers.

Example 2-7: Definitions of reference **r** and pointer **ipt**.

//--
// File: **example2_7.cpp**
// This program demonstrates the difference between reference r and pointer ipt.
//--
```
1   #include <iostream>
2   using namespace std;
3
4   int main()
5   {
6       int i1;
7       int& r = i1;         // a reference
8       int* ipt = &i1;      // a pointer
```

```
9    i1 = 5;
10   cout << "r=" << r <<endl;
11   cout << "&i1=" << &i1 <<endl;
12   cout << "&r=" << &r <<endl;
13   cout << "*ipt=" << *ipt <<endl;
14   cout << "ipt=" <<ipt<<endl;
15   cout << "&ipt=" <<&ipt<<endl;
16   cout<<"-----------------------\n";
17   int i2 = 8;
18   r = i2;
19   ipt = &i2;
20   cout << "r=" << r <<endl;
21   cout << "&i1=" << &i1 <<endl;
22   cout << "&i2=" << &i2 <<endl;
23   cout << "&r=" << &r <<endl;
24   cout << "*ipt=" << *ipt <<endl;
25   cout << "ipt=" << ipt <<endl;
26   cout << "&ipt=" << &ipt <<endl;
27
28   return 0;
29 }
```

Result:

```
1    r=5
2    &i1=0012FF7C
3    &r=0012FF7C
4    *ipt=5
5    ipt=0012FF7C
6    &ipt=0012FF74
7    -----------------------
8    r=8
9    &i1=0012FF7C
10   &i2=0012FF70
11   &r=0012FF7C
12   *ipt=8
13   ipt=0012FF70
14   &ipt=0012FF74
```

In this example, **r** is of type "reference to *int*", and **ipt** is of type "pointer to *int*". The value of reference **r** is an integer, but the value of pointer **ipt** is a memory address, Figure 2-2 shows their values in the memory.

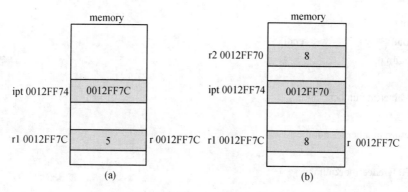

Figure 2 2 Variables **r** and **ipt** in Example 2-7
(a) Definition of variables r and ipt; (b) After statements r = r2 and ipt = &r2.

2.5.2 Reference Variables as Parameters

Parameters provide a communication link between the calling function (such as *main*) and the called function. They enable functions to manipulate different data each time they are called. In general, there are two types of formal parameters: **value parameters** and **reference parameters**.

> **Value parameter** is a formal parameter that receives a copy of the content of the corresponding actual parameter.
>
> **Reference parameter** is a formal parameter that receives the location (memory address) of the corresponding actual parameter.

The following program shows how a value parameter and a reference parameter work.

Example 2-8: Reference variables as parameters.

```
//----------------------------------------------------------------
// File: example2_8.cpp
// This program defines two functions, having a reference parameter and a value parameter.
//----------------------------------------------------------------
1   #include <iostream>
2   using namespace std;
3
4   int byValue( int );              // passing arguments by value
5   int byReference( int& );         // passing arguments by reference
6
7   int main()
8   {
9       int x = 2;
10      int y = 4;
11      cout << byValue( x ) << endl;
12      cout << "x = " << x << endl;
```

```
13        cout<< byReference( y );
14        cout << "y = " << y << endl;
15        return 0;
16   }
17   int byValue(int number)
18   {
19        return number *= number;
20   }
21   int byReference(int &number)
22   {
23        return number *= number;
24   }
```
Result:
1   4
2   x = 2
3   16
4   y = 16

The statement in Line 11 calls the function *byValue*. The value of the actual parameter **x** is then passed to a formal parameter **number**. After the statement executes, the function *byValue* returns the value of **number *= number**. When the function executes, any changes made to the formal parameter **number** do not in any way affect the actual parameter **x**. The actual parameter **x** have no knowledge of what is happening to the formal parameter **number**. So the statement in Line 12 outputs 2 for **x**. Thus, value parameters provide a one-way link between actual parameters and formal parameters. Hence, functions with value parameters have limitations.

The statement in Line 13 calls the function *byReference*. The value of the actual parameter is passed to a formal parameter **number**. Likewise, the function *byReference* returns the value of **number *= number** after the statement executes. Because the formal parameter **number** is a reference parameter, it receives the address (memory location) of the actual parameter **y**. Thus, reference parameter **number** can pass one value for the function and can change the value of the actual parameter **y**. So the statement in Line 14 outputs 16 for **y**.

Reference parameters are useful in three situations:
- when you want to return more than one value from a function
- when the value of the actual parameter needs to be changed
- when passing the address would save memory space and time relative to copying a large amount of data

2.5.3 References as Value-Returning

Returning values from a function to its caller by value, address, or reference works almost exactly the same as passing parameters to a function. They have the same advantages and disadvantages. The primary difference between the two is simply the opposite directions of data flow. However, there is one more added bit of complexity — because local variables in a function go out of scope when a function returns, we need to consider the effect of this on each return type.

Just like pass by reference, values returned by reference must be variables. When a variable is returned by reference, a reference to the variable is passed back to the caller. The caller can then use this reference to continue modifying the variable, which can be useful at times. Return by reference is also fast, which becomes useful when returning structs and classes.

Example 2-9: References as return values.

```
//---------------------------------------------------------------------------------
// File: example2_9.cpp
// This program defines two functions which return by reference and value.
// ---------------------------------------------------------------------------------
1    #include <iostream>
2    using namespace std;
3
4    int returnValue(int n)
5    {
6        int nValue = n * 2;
7        return nValue;          // A copy of nValue will be returned here
8    }   // nValue goes out of scope here
9
10   int& returnReference(int n)
11   {
12       int nValue = n * 2;
13       return nValue;          // return a reference to nValue here
14   }   // nValue goes out of scope here
15
16   int main()
17   {
18       int   x = 4;
19       cout << " returnValue = " << returnValue(x) << endl;
20       cout << "returnReference = " << returnReference(x) <<endl;
21       return 0;
22   }
```

Result:
```
1    returnValue = 8
2    returnReference = 8
```

- Return by value is the simplest and safest return type to use. When a value is returned by value, a copy of that value is returned to the caller.
- As with pass by value, you can return by value literals (e.g. 5), variables (e.g. x), or expressions (e.g. x+1), which makes return by value very flexible.
- As with pass by reference, you can not return a reference to a literal or expression.
- As with return by value, you can return a variable (or expression) that involves local variable **nValue** declared within function *returnValue*. Because the variable is evaluated before the function goes out of scope, and a copy of **nValue** is returned to the caller, there are no problems when the variable goes out of scope at the end of the function.
- As with return by reference, the problem within the function *returnReference* — is trying to return value **nValue**, which goes out of scope when the function returns, passed by reference to the function and back to called. To solve this problem, we can declare **nValue** to be a global variable.

2.5.4 References as Left-Hand Values

Return by reference is typically used to return arguments passed by reference to the function back to the caller—used on the right-hand side of an assignment. If the function call can appear on the left hand side of an assignment, what will happen? In the following example, we return (by reference) an element of an array that was passed to our function by reference on the left-hand of an assignment.

A **left-hand value** is the value on the left-hand side of an assignment operation.

Example 2-10: Reference as a left-hand value.

//---
// File: **example2_10.cpp**
// This program assigns a value to the element of an array using left-hand value.
//---

```
1   #include <iostream>
2   using namespace std;
3
4   int a[10];
5
6   int & Arr(int i);
7   {   return a[i];   }
8
9   int main()
10  {
11      Arr(4) = 50;      ///left-hand
12      cout << a[4];
```

```
13        return 0;
14   }
```
Result:
```
1   50
```

When we call *Arr(4)* in Line11, function *Arr* returns a reference to the 4th element of the array **a**. The *main* function then uses this reference to assign the 4th element the value 50.

Although this is somewhat of a contrived example (because you could access **a** directly), once you learn about classes, you will find a lot more uses for returning values by reference.

Think These Over

1. What are the differences between *pointers* and *references*?
2. What is a principal reason for passing parameters by reference?

2.6 Namespaces

A **namespace** is a mechanism for expressing logical grouping, i.e. if declarations logically belong together according to specific criteria, then a common namespace can be used to express this fact.

The syntax of the statement **namespaces** is

namespace identifier
{
    entities
}

where **identifier** is any valid identifier and entities is the set of classes, objects and functions that are included within the namespace. For example,

```
namespace mySpace {
      int a, b;
      double f, d;
}

namespace yourSpace {
      int a, b;
      double f, d;
}
```

In this example, variables **a**, **b**, **f** and **d** are declared within two namespaces called *mySpace* and *yourSpace* respectively. Although these variables have the same name, declaring them within the two namespaces will not cause a redefinition error; the namespace is able to avoid **name collisions** (of variables, types, classes or functions).

In order to access these variable from the *mySpace* and *yourSpace* namespaces, we have to use the scope operator **::**. For example, to access variable **a** from outside the namespaces, we can write, mySpace::a

yourSpace::a

Example 2-11: Definition of two namespaces.

//--

// File: **example2_11.cpp**

// This program defines two namespaces, mySpace and yourSpace.

//--

```
1   #include <iostream>
2   using namespace std;
3
4   namespace mySpace
5   {
6       int var = 5;
7   }
8
9   namespace yourSpace
10  {
11      double var = 3.1416;
12  }
13
14  int main()
15  {
16      cout << mySpace::var << endl;
17      cout << yourSpace::var << endl;
18      return 0;
19  }
```

Result:

1   5

2   3.1416

The keyword **using** is used to introduce a name from a namespace into the current declarative region.

Example 2-12: Utilizing the *using* keyword for namespaces.

//--

// File: **example2_12.cpp**

// This program utilizes the using keyword for namespaces.

//--

```
1 #include <iostream>
2   using namespace std;
3   namespace mySpace
4   {
```

```
5        int a = 5;
6        int b = 6;
7    }
8    namespace yourSpace
9    {
10       double a = 3.1416;
11       double b = 5.678;
12   }
13   int main()
14   {
15       using mySpace::a;
16       using yourSpace::b;
17       cout << a << endl;
18       cout << b << endl;
19       cout  <<   mySpace::b << endl;
20       cout  <<   yourSpace::a << endl;
21       return 0;
22   }
```

Result:

```
1    5
2    5.678
3    6
4    3.1416
```

Notice how in this example, variable **a** (without any name qualifier) refers to **mySpace::a** in Line 17, whereas variable **b** refers to **yourSpace::b** in Line 18, exactly as our using declarations have specified. We still have access to **mySpace::b** and **yourSpace::a** using their fully qualified names.

Example 2-13: The same name function in namespaces.

```
//-----------------------------------------------------------------------------------
// File: example2_13.cpp
// This program defines the same name variables and functions within two namespaces.
//-----------------------------------------------------------------------------------
1    #include <iostream>
2    using namespace std;
3
4    namespace mySpace
5    {
6        int a = 5;
7        int b = 6;
8        int add(int, int);
9    }
10
```

```
11    namespace yourSpace
12    {
13          double a = 3.1416;
14          double b = 5.678;
15          double add(double, double);
16    }
17
18    int mySpace::add(int x, int y) { return x + y; }
19    double yourSpace::add(double x, double y)    { return x + y; }
20
21    int main()
22    {
23          using namespace mySpace;
24          cout << a << endl;
25          cout << b << endl;
26          cout << add(a, b);
27
28          std::cout << yourSpace::a << endl;
29          std::cout << yourSpace::b << endl;
30          std::cout << yourSpace::add (yourSpace::a, yourSpace::b) << endl;
31
32          return 0;
33    }
```

Result:

1 5
2 6
3 11
4 3.1416
5 5.678
6 8.8196

In this example, we define the function **add** within the two namespaces respectively. Since we have declared that we were using the statement *using namespace mySpace* in Line 23, all direct uses of the variables **a** and **b**, and the function **add** without name qualifiers (in Lines 24, 25 and 26) are referring to their declarations in namespace *mySpace*. So we can do for the variables and functions within the namespace *yourSpace*.

Here the function bodies are defined outside the namespaces mySpace and yourSpace, the aim of this is to separate the interface (i.e. function declaration) from the implementation (i.e. function definition).

As a result of separating the implementation of the interface, each function now has exactly one declaration and one definition. Users will see only the interface containing declarations. The implementation — in this case, the function bodies — will be placed "somewhere else" where a user need not look.

You may notice the **std** specifier is used in Example 2-13. Because all the files in the C++ standard library declare all of its entities within the std namespace, we have generally included the *using namespace std*; statement in all programs that used any entity defined in iostream.

Word Tips

allocate *vt.* 分配	hint *vt.* 暗示，示意
arguments *n.* 自变量	identical *adj.* 同一的，相同的
array *n.* 数组	identifiers *n.* 标识符
actual paramenter 实参	illustration *n.* 举例说明
brackets *n.* 括弧	implementation *n.* 实现
built-in *adj.* 系统建立的	implicit *adj.* 隐含的
comments *n.* 注释	interface *n.* 接口，界面
compatibility *n.* 兼容性	linkage *n.* 链接
comprise *vt.* 包含	manipulates *v.* 计算，处理
contexts *n.* 上下文	mechanism *n.* 机制
contrived *adj.* 不自然的，做作的	multiple *adj.* 许多的
corresponding *v.* 对应，相配	notion *n.* 概念
criteria *n.* 标准	optionally *adv.* 任意地
curly braces 花括号	parentheses *n.* 圆括号
decimal *adj.* 小数的，十进制的	presumably *adv.* 大概
elaborate *vi.* 变复杂的	prototypes *n.* 原型
embedded *adj.* 嵌入的	semantics *n.* 语义学，语意论
enum *n.* 枚举	substituted *adj.* 代替的
evaluated *vi.* 赋值，评估	syntax *n.* 语法，句法
execution *n.* 执行	trailing *adj.* 牵引的，后面的
extensive *adj.* 广泛的	variables *n.* 变量
flexibility *n.* 灵活性	volatile *adj.* 易变的
formal parameter 形参	

Exercises

1. Answer the following questions:
(1) Dividing a program into functions
   a. is the key to object-oriented programming.
   b. makes the program easier to conceptualize.
   c. may reduce the size of the program.
   d. makes the program run faster.
(2) A function's single most important role is to
   a. give a name to a block of code.
   b. reduce program size.
   c. accept arguments and provide a return value.

Object-Oriented Programming in C++

  d. help organize a program into conceptual units.

(3) A default argument has a value that

  a. may be supplied by the calling program.

  b. may be supplied by the function.

  c. must have a constant value.

  d. must have a variable value.

(4) A function argument is

  a. a variable in the function that receives a value from the calling program.

  b. a way that functions resist accepting the calling program's values.

  c. a value sent to the function by the calling program.

  d. a value returned by the function to the calling program.

(5) Overloaded functions

  a. are a group of functions with the same name.

  b. all have the same number and types of arguments.

  c. make life simpler for programmers.

  d. may fail unexpectedly due to stress.

(6) When an argument is passed by reference

  a. a variable is created in the function to hold the argument's value.

  b. the function cannot access the argument's value.

  c. a temporary variable is created in the calling program to hold the argument's value.

  d. the function accesses the argument's original value in the calling program.

2. Read the following program and finish it:

```
#include <iostream>
using namespace std;
    int& put(int n);    //put value into the array
    int get(int n);     //obtain a value from the array
    int vals[10];
    int error=-1;
    void main()
    {
        put(0)=10;      // put value into the array
        put(1)=20; put(2)=30;
        cout<<get(0)<<endl;
        cout<<get(1)<<endl;
        cout<<get(2)<<endl;
        put(12)=1;      //out of range
    }
```

3. Write out the output of the following programs.

(1)

```
int i = 7;
int& r = i;
```

```
        r =9;
        i = 10;
        cout<< r<<" ; "<< i<<endl;
        cout<<&r<<":"<< &i<<endl;
```
(2)
```
        namespace mySpace{
            int a = 5, b = 6;
        }
        namespace yourSpace{
            double a = 4.5, b = 98.3;
        }
        void main()
        {
            using mySpace::a ;
            using yourSpace::b;
            cout<<a<<";"<<b<<endl;
        }
```
(3)
```
        #include <iostream>
        using namespace std;
        void Test(int&, int);
        void main()
        {
            int x = 12;
            int y = 14;
            Test(x, y);
            cout<<"After the first call of Test, the variables equal " << x << y;
            x = 20;
            y = 22;
            Test(x, y);
            cout<<"After the second call of Test, the variables equal " << x << y;
        }
        void Test(int& a, int b)
        {
            a = 3;
            a += 2;
            b = a *5;
            cout<<"In function Test, the variable equal "<<a<<b<<endl;
        }
```
4. Write a function that swaps two integer numbers and two double floating-point numbers. Hint: Using *reference and pointer* as the parameters.
5. Write a program that has three overloading function *display* to produce them output. The first

function returns a *double* type. The second function returns an *int* type. The third function returns a *char* type.

6. Write a function that takes an integer value and return the number with its digits reversed. For example, given the number 1234, the function should return 4321.

7. Write a function **qualityPoints** that inputs a student's average and a returning value - 4 if a student's average is 90-100, 3 if the average is 80-89; 2 if the average is 70-79, 1 if the average is 60-69, and 0 if the average is lower than 60.

8. Write a function that finds the maximum and minimum values of an array.

9. Write a function *integerPower (base, exponent)* that returns the value of base exponent. For example, integerPower(3, 4)=3*3*3*3. Assume the exponent is a positive, nonzero integer and that base is an integer. (Do not use any math library function)

10. Write a program that prompts the user for gallons, and change it to liters and prints the liters. (1 gallon=3.78533liter) The program should have a const representing the value of gallon changed to liter.

11. Write a program that inputs 10 integer numbers and outputs the sum of numbers (>=0) and the sum of numbers (<0).

12. Write a program to calculate the calorie of lunch. User input three lines. Each line has a character (M is for meat, V is for vegetable, D is for dessert) and calorie value. The program outputs the total calorie value of lunch.

13. According to input string, statistics the number of digitals and alphabets and output them.

14. Write a program to solve roots of $ax^2+bx+c=0$ equation.Hint:for different a, b and c, the equation may have 0, 1 or 2 roots. You may define a function as follows:
    int *solving*(double a, double b, double c, double& x1, double& x2)

where parameters x1 and x2 are the roots. The returning value of the function is the number of roots.

15. Write a program that plays the game "guess the number" as follows. Your program chooses the number to be guessed by selecting an integer at random in the range 1 to 1000. The program then displays the following:

"I have a number between 1 and 1000. Can you guess my number?

Please type your number:"

The player then types a guess. The program responds with one of the following after every guess:

(1) Bingo! You guessed the number.
(2) Too high. Try again…
(3) Too low. Try again...

And the program count the number of guesses that the player makes.

(1) If the number is 5 or fewer, print "You know the secret?"
(2) If the player guesses the number in 10 tries, then print "You got the lucky!"
(3) If the player makes more than 10 guess, then print "You should be able to be better."

Chapter 3
Classes and Objects (I)
—*Data Abstraction and Definition of Classes* —

> *Private faces in public places are wiser and*
> *nicer than public faces in private places.*
> —*W. H. Auden*

Objectives
- To understand data abstraction and encapsulation
- To be able to create a user-defined type, namely class
- To understand how classes are implemented
- To be able to define constructors and destructors
- To be able to use objects

3.1 Structures

A structure type is a user-defined type. Structures are aggregate data type built using elements of other types, that is, structures are a way of storing many different values in variables of potentially different types under the same name. This makes it a more modular program, which is easier to modify because its design makes things more compact. For example, a structure can be used to store information about students in a database.

When you define a structure, you must remember that the main difference between structures and arrays is that the structure is a collection of different data types and the array is a collection of the same data type.

3.1.1 Defining a Structure

The structure is defined by using the keyword **struct** followed by structure name. Consider the following structure definition:

```
struct Date{
        int day;        // 1-31
        int month;      // 1-12
        int year;       // 1-9999
    };
```

The keyword **struct** introduces the structure definition. The identifier *Date* is the structure tag that names the structure definition and is used to declare variables of the structure type. In this example, the new type name is **Date**. The names declared in the braces of structure definition are the structure's **members**. Members of the same structure must have unique names, but two different structures may contain members of the same name without conflict. Each structure definition must end with semicolon.

The preceding structure definition does not reserve any space in memory, rather, the definition creates a new data type that is used to declare variable. Structure variable are declared like variables of other types. For example, the declaration of structure **Date** is

   Date dateObject, dateArray[4];

3.1.2 Accessing Members of Structures

To access an individual member of a struct variable, you can use the *member access operators*, the *dot operator* (.) and the *arrow operator* (->). For example, to output member **day** of structure **dateObject** uses the following statement

   Date dateObject;

   cout<< dateObject.day;

Assume that the pointer **dataPtr** has been declared to point to a **Date** object, and that the address of structure **dateObject** has been assigned to **datePtr**. To output member **day** of structure **dateObject** with pointer **datePtr**, uses the following statement

   date* datePtr = &dateObject;

   cout << datePtr ->day;

Implementing a User-Defined Type Date with a Struct

The program defines a single **Date** structure called **today** and uses the dot operator to initialize the structure members with the values 18 for **day**, 11 for **month**, and 2010 for **year**. The program then adds a number to member **year** by calling function *add_year*, and prints the date in the English standard format.

Example 3-1: A user-defined type **Date** with a struct.

//---

// File: **example3_1.cpp**

// This program defines a structured type **Date** and declares a structured variable **today**.

// The program adds a number to the year of **today** and prints the date by calling two functions

// that are defined outside of the structured type.

//---

```
1    #include <iostream>
2    using namespace std;
3    //Defining a structure
4    struct Date {
5        int day;
```

```
6          int month;
7          int year;
8      };
9   //Defining functions for manipulating memberrs of the structure
10  void init_date(Date& date, int d, int m, int y);
11  void add_year(Date& date, int n);
12  void print(Date& date);
13
14    int main()
15    {
16         Date today;                        //declaring a structure type
17
18         init_date(today, 18, 11, 2010);    //set members to valid values
19
20         cout<<" The date of today is : ";
21         print(today);                      //print today
22         cout<<endl;
23
24         add_year(today, 2);                // add 2 to year
25         cout<<" The date of today is : ";
26         print(today);                      // print today
27         return 0;
28    }
29
30  //Initialize a date
31  void init_date (Date& date, int d, int m, int y)
32  {
33       date.day = d;    date.month = m;    date.year = y;
34  }
35  // Add a number to member year
36  void add_year(Date& date, int n)
37  {
38         date.year += n;
39  }
40  //Print a date in the english standard format
41  void print(Date& date)
42  {
43         cout<< date.month <<"-"<< date.day << "-" << date.year;
44  }
```

Result:

1 The date of today is :11-18-2010
2 The date of today is :11-18-2012

In this example, the **Date** structured type contains three members of integer types, namely, **day, month** and **year**. When you declare a variable **today** of type **Date**, **today** does not represent just one data value of the member for type **Date**, it represents an entire collection of characters. Each of the members in **today** can be accessed individually, as the statement in Line 38 or 43 described. You may notice that three functions in Lines 30, 35 and 40 are defined for manipulating the members of **Date**.

Think It Over

Why are the functions passed by a reference parameter of type **Date** in Example 3-1?

3.1.3 Structures with Member Functions

In Example 3-1, we used the structure to define a date with variables and a set of functions for manipulating variables on this type. However, there is no explicit connection between the data type and these functions. Such a connection can be established by declaring the functions as members inside the structure. Thus, Example 3-1 can be changed into the following form.

```
1   #include <iostream>
2   using namespace std;
3   //Defining a structure
4   struct Date {
5        int day;
6        int month;
7        int year;
8
9        void init_date(int d, int m, int y);
10       void add_year(int n);
11       void print();
12  };
13
14  int main()
15  {
16       Date today;                        //declaring a structure type
17
18       today.init_date(18, 11, 2010);     //set members to valid values
19
20       cout<<" The date of today is : ";
21       today.print();                     //print today
22       cout<<endl;
23
24       today.add_year(2);                 // add 2 to year
```

52

```
25      cout<<" The date of today is : ";
26      today.print();                    // print today
27      return 0;
28  }
29
30  //Initialize a date
31  void Date::init_date (int d, int m, int y)
32  {
33      day = d;    month = m;   year = y;
34  }
35  // Add a number to member year
36   void Date::add_year(int n)
37   {
38      year += n;
39   }
40  //Print a date in the english standard format
41   void Date::print()
42   {
43      cout<< month <<"-"<< day << "-" << year;
44   }
```

Result:

1 The date of today is :11-18-2010
2 The date of today is :11-18-2012

The result of this program is the same as one of Example 3-1. However, the functions defined in Lines 9, 10 and 11 are placed inside the **Date** structure as members. These functions are accessed using a member access operator (.), as mentioned in Lines 18, and 21.

3.2 Data Abstraction and Classes

3.2.1 Data Abstraction

As the developed software becomes more complex, we design methods and data structures in parallel. We progress from the logical or abstract data structure envisioned at the top level through the refinement process until we reach the concrete coding.

Let's take an example of TV which you can turn on and off, change the channel and adjust the volume. You know how to use a TV. But you do not know it's internal detail that is, you do not know how it receives signals over the air or through a cable, how it translates them, and finally displays them on the screen.

Separating the design detail (that is, how the TV's works) from its use is called abstraction. In the words, abstraction focuses on what a TV does and not on how it works.

> **Data abstraction** is a process of separating the logical properties from the implementation detail. It is an OO methodology.

In this example, turning on a TV is a logical property. The internal process of turning on constitutes the implementation detail. The user is not interested in this process.

Abstraction can also be applied to data. We presented a Date structured type in Example 3-1. Now, let us illustrate data abstraction using the Date type. Definition of a Date type is as follows:

1. Domain

Each Date value is a date in the form of year, month and day, that is, the Date type has three member variables, i.e. year, month and day.

2. Operations

(1) Initialize a date (e.g. void init_date(int d, int m, int y));

(2) Add a number to the year (e.g. void add_year(int n));

(3) Print a date in the English standard format (e.g. void print()).

The actual implementation of these functions manipulating on the member variables is separated.

3.2.2 Defining Classes

Chapter 1 introduced the programming methodology called object-oriented programming (OOP). In OOP, the first step is to identify the components (called objects) through using the principle of data abstraction. An object is the variable of a class which combines data and the operations on that data in a single unit. The aim of the C++ class concept is to provide the programmer with a tool for creating new types that can be used as conveniently as the built-in types.

A class is a user-defined type. Classes enable the programmer to model objects that have *attributes* (represented as *data members*) and *behaviours* or *operations* (represented as *member functions*). Types containing data members and member functions are defined in C++ using the keyword **class**.

Member functions are sometimes called ***methods*** in other object-oriented programming languages, and are invoked in response to *message* sent to an object. A message corresponds to a member-function call sent from one object to another or sent from a function to an object.

> A **class** is a user-defined type. A **class** is a set or collection of abstracted objects that share common characteristics. It consists of both data values and operations on those values.

Classes are generally declared using the keyword ***class***, with the following format:

```
class class_name {
        access_specifier1:
                // members;
        access_specifier2:
                //members;  };
```

where `class_name` is a valid identifier for the class. The body of the declaration within braces can contain members, which can be either data or function declarations, and optionally access specifiers (see §3.4). Like a structure, the body of the class is delimited by braces ({}) and terminated by a semicolon (;). (**Do not forget the semicolon**. Remember, data constructs, such as structures and classes, end by using a semicolon, whereas control constructs, such as functions and loops, do not.)

> The components of a class are called the *members of the class*. The members of the class may be either data type or functions. Data in the class are called *data members*. Functions declared within the class definition are called *member functions*.

For example,

```
Class Date{
    private:                              // access control specifier
        int day;                          // 1-31
        int month;                        // 1-12
        int year;                         // 1-9999
    public:                               // access control specifier
        void init_date (int d, int m, int y);   //initialize
        void add_year ( int n);           // add n years
        void print();                     // print a date
};
```

In this example, integer variables day, month and year are the *data members* of class Date; functions *init_date*, *add_year* and *print* are its *member functions*; the *private* and *public* keywords are access specifier.

3.2.3 Defining Objects

Once a class has been defined, the class name can be used to declare objects of that class.

> An **object** is an instance of a class or a variable of a class.

An object does not exist until an instance of the class has been created; the class is just a definition. When the object is physically created, the space for that object is allocated in the memory. It is possible to have multiple objects created from one class.

Objects are declared like variable of other types. The declaration

    Date today, tomorrow;
    Date *day;

creates three objects of the **Date** class:day *today* and *tomorrow*. Each object has its own copies of **day**, **month** and **year**, the private data members of the class.

3.2.4 Accessing Member Functions

Like a structure variable, members of an object are accessed using the *member access operators* — the *dot operator* (.) and the *arrow operator* (->).For example,

```
today.init_date(18,11,2010);
tomorrow.init_date(19,11,2010);
day->init_date(23, 8, 2003);
```

At a given moment during program execution, data members **day**, **month** and **year** of the **today** object are assigned with the values 18, 11 and 2010 by invoking function *init_date*, whilst tomorrow's data members are assigned with the values 19, 11 and 2010.

Figure 3-1 shows a visual image of the class objects **today** and **tomorrow**.

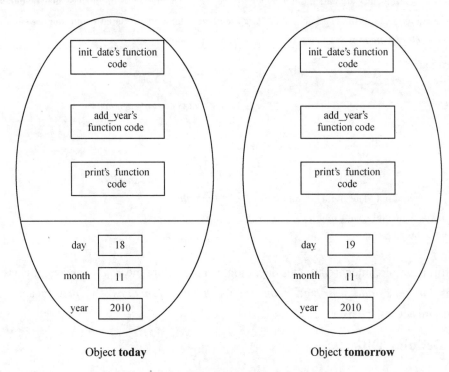

Figure 3-1   Conceptual view of two objects

Implementing a Date Abstract Data Type with a Class

Data abstraction is defined as a process of separating the logical properties of the data from its implementation. Every data type consists of a set of values (data members) along with a collection of operations (member functions) on those values. For example, the data members (e.g. **year**, **month** and **day**) of class **Date** and its basic operations are the logical properties; the algorithms to perform these operations are the implementation detail of **Date**.

> **Abstract data type (ADT)** — a data type that separates the logical properties from the implementation detail.

Example 3-2: An abstract data type — class **Date**.

```
//---------------------------------------------------------------------------
// File: example3_2.cpp
```

// This program defines a Date class type and creates Date objects.

```cpp
//------------------------------------------------------------------------
1    #include <iostream>
2    using namespace std;
3    class Date {            //class definition
4        int day;
5        int month;
6        int year;
7    public:
8        //Initialize a date
9        void init_date ( int d, int m, int y)
10       {
11           day = d;   month = m;   year = y;
12       }
13       // Add a number to member year
14       void add_year ( int n)
15       {
16           year += n;
17       }
18       //Print a date in the English standard format
19       void print ()
20       {
21           cout<< month <<"-"<<day << "-" <<year;
22       }
23   };
24
25   int main()
26   {
27       Date today, tomorrow;            // create two objects
28
29       today. init_date( 18, 11, 2010);   //set members to valid values
30       tomorrow.init_date( 19, 11, 2010);
31
32       cout<<" The date of today is : ";
33       today.print();
34       cout<<endl;
35
36       cout<<" The date of tomorrow is : ";
37       tomorrow.print();
38       cout<<endl;
39
40       today.add_year( 2);       //add a number to year
```

```
41          tomorrow.add_year(2);
42
43          cout<<" The new date of today is : ";
44          today.print();
45          cout<<endl;
46          cout<<" The new date of today is : ";
47          tomorrow.print();
48          cout<<endl;
49          return 0;
50    }
```

Result:

```
1  The date of today is :11-18-2010
2  The date of tomorrow is :11-19-2010
3  The new date of today is :11-18-2012
4  The new date of tomorrow is : 11-19-2012
```

In a member function, member names can be used without explicit reference to an object for which the function was invoked. For example, function *init_date* is invoked by objects **today** and **tomorrow** respectively.

3.2.5 In-Class Member Function Definition

A member function defined within the class definition — rather than simply declared there — is taken to be an **inline** member function. That is, in-class definition of member functions is for small, frequently-used functions. Consider

```
1   //Date.cpp
2   class Date {
3   public:
4       void init_date(int, int, int);
5       void add_year(int n)
6       { year += n;   }
7       void print();
8   private:
9       int day, month, year;
10  };
```

The statements in Lines 5 and 6 state that the member function *add_year* is defined inside the class. The other functions are defined outside the class. This is perfectly good C++ code because a member function declared within a class can refer to every member of the class as if the class were completely defined before the member function bodies were considered. However, this can confuse human readers.

Consequently, we usually either place the data first or define the inline member functions after the class itself. For example,

```
1  //Date.cpp
2  class Date {
3  public:
4      void init_date(int, int, int);
5      void add_year(int );
6      void print();
7  private:
8      int day, month, year;
9  };
10 inline void Date::add_year(int n)
11 { year += n; }
```

Member Functions Defined Outside the Class

We have seen member functions that were defined inside the class definition in Example 3-2. This need not always be the case. The statement in Line 11 of the above example shows a member function *add_year* that is not defined within the **Date** class definition. It is only *declared* inside the class, with the statement

void add_year(int);

This tells the compiler that this function is a member of the class but that it will be defined outside the class declaration, someplace else in the listing.

In this situation, the member function is defined in the following syntax.

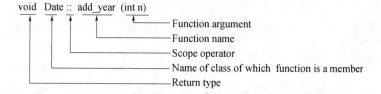

The Date class in Example 3-2 can also be defined as follows:

```
1 //Date.cpp
2 class Date {
3 public:
4     void init_date(int, int, int);
5     void add_year(int);
6     void print();
7 private:
8     int day, month, year;
9 };
10 void Date::init_date(int d, int m, int y)
11 {   day = d; month = m; year = y;   }
```

```
12 void Date::add_year(int n)
13 {   year += n;   }
14 void Date:: print()
15 {    cout << month <<"-"<<day << "-" <<year;   }
```

3.2.6 File Structure of an Abstract Data Type

An abstract data type (ADT) consists of two parts: a specification and an implementation. The specification describes the behaviour of the data type without reference to its implementation. The implementation creates an abstraction barrier by hiding the concrete data representation as well as the code for the operations.

Normally, the **Date** class declaration servers as the specification of Date. It is described in a specification file which is called **header file** (or **interface file**).

Example 3-3:   File structure of the **Date** abstract data type.

//--
// File: **Date.h** (specification file)
// This program describes the definition of a Date abstract date type.
//--
```
1     class Date {
2     public:
3          void init_date(int, int, int);
4          void add_year (int);
5          void print();
6     private:
7          int day;
8          int month;
9          int year;
10    };
```

The specification file (.h) for the **Date** class contains only the class declaration. The implementation file (.cpp) must provide the definitions of all the class member functions.

//--
// File: **Date.cpp** (implementation file)
// The program presents the implementation detail of the Date class.
//--
```
1     #include   "Date.h"              //referring the declaration of the class Date
2     #include <iostream>
3
4     using namespace std;
5
6     //Initialize a date
7     void Date::init_date ( int d, int m, int y)
8     {
```

```
9          day = d;   month = m;   year = y;
10   }
11   // Add a number to member year
12   void Date::add_year ( int n )
13   {
14          year += n;
15   }
16   //Print a date in the English standard format
17   void Date::print ()
18   {
19          cout<< month <<"-"<<day << "-" <<year;
20   }
```

The most important new thing in this code is the operator of scope (::, two colons) included in the definition of three member functions *init_date, add_year* and *print*. They are used to define a member of a class from outside the class definition itself.

The scope operator (::) specifies the class to which the member being declared belongs, granting exactly the same scope properties as if this function definition was directly included within the class definition (compared with Example 3-2).

For example, in the function *init_date* of the previous code, we have been able to use the variables **day**, **month** and **year**, which are private members of class **Date**, which means they are only accessible from other members of their class.

You should have a user file (.cpp) to use the **Date** class.

```
//----------------------------------------------------------------
// File: example3_3.cpp (user file)
// This program creates Date objects and accesses their members.
//----------------------------------------------------------------
1   #include "Date.h"              //referring the declaration of the class
2   #include <iostream>
3
4   using namespace std;
5   int main()
6   {
7          Date today, tomorrow;        // create two objects
8
9          today. init_date( 18, 11, 2010);    //set members to valid values
10         tomorrow.init_date( 19, 11, 2010);
11
12         cout<<" The date of today is : ";
13         today.print();
14         cout<<endl;
15
```

```
16      cout<<" The date of tomorrow is : ";
17      tomorrow.print();
18      cout<<endl;
19
20      today.add_year( 2);           //add a number to year
21      tomorrow.add_year(2);
22
23      cout<<" The new date of today is : ";
24      today.print();
25      cout<<endl;
26      cout<<" The new date of today is : ";
27      tomorrow.print();
28      cout<<endl;
29      return 0;
30   }
```

Result:

1   The date of today is :11-18-2010
2   The date of tomorrow is :11-19-2010
3   The new date of today is :11-18-2012
4   The new date of tomorrow is : 11-19-2012

3.3 Information Hiding

The previous section defined the class **Date** to implement the date in a program. We then wrote a program (Date.cpp) that used the class **Date**. In fact, we combined the class **Date** with the function definitions to implement the operations and the function *main* so as to complete the program. That is, the specification and implementation details of the class **Date** were directly incorporated into the program.

Is it a good practice to include the specification and implementation detail of a class in the program? Definitely not. There are several reasons for not doing so.

Suppose the definition of the class and the definitions of the member functions are directly included in the user's program (see Example 3-2). The user then has direct access to the definition of the class and the definitions of the member functions. Therefore, the user can modify the operations in any way the user please. The user can also modify the data members of an object in any way the user please. Thus, in this sense, *the **private** data members of an object are no longer **private** to the object.*

If several programmers use the same object in a project, and if they have direct access to the internal parts of the object, there is no guarantee that every programmer will use the same object in exactly the same way. Thus, we must hide the implementation details. *The user should know only what the object does, not how it does it.*

Hiding the implementation details frees the user from having to fit this extra piece of code in the program. Also, by hiding the details, we can ensure that an object will be used in exactly the same way throughout the project. Furthermore, once an object has been written, debugged, and tested properly, it becomes (and remains) error-free.

To implement **Date** in a program, the user must declare objects of class Date, and know while operations are allowed and what the operations do. So the user must have access to the specification detail. Because the user is not concerned with the implementation details, we must put those details in a separate file (e.g. Date.cpp in Example 3-3).

Additionally, we must free the user from having to including them directly in the program because the specification details can be too long.

For example, in the **example3_3.cpp** program of Example 3-3, the user can do:
#include "Date.h"

Notice that the user must be able to look at the specification details so that he or she can correctly call the functions, and so forth.

Think These Over

1. Why do we need to use data abstraction in OOP?
2. What is an abstract data type?
3. What is the purpose of information hiding?

3.4 Access Control

All is very similar to the declaration on data structures, except that we can not include only functions and members, but also this new thing called *access specifier*. An access specifier is one of the following three keywords: ***private***, ***public*** or ***protected***. These specifiers modify the access rights that the members following them acquire:

• *private* members of a class are accessible only from within other members of the same class and friends (see Section4.8).

• *protected* members are accessible from members of their same class and from their friends, but also from members of their derived classes. (see Section 6.5.1)

• *public* members are accessible from anywhere where the object is visible.

The body of the class contains two unfamiliar keywords: *private* and *public*. What is their purpose?

A key feature of object-oriented programming is *information (data) hiding*. This term does not refer to the activities of particularly paranoid programmers; rather it means that data is concealed within a class so that it cannot be accessed mistakenly by functions outside the class. The primary mechanism for hiding data is to put it in a class and make it private. Private data or functions can only be accessed from within the class. Public data or functions, on the other hand, are accessible from outside the class.

The declaration of **Date** in the previous section provides a set of functions for manipulating a date. However, it does not specify that those functions should be the only ones to depend directly on **Date**'s representation and the only ones to directly access objects of class **Date**. This restriction can be expressed by using a *class* instead of a *struct*:

```
class  Date{
private:
          int day;         // 1-31
          int month;       // 1-12
          int year;        // 1-9999
    public:                // access control specifier
          void init_date (int d, int m, int y);   //initialize
          void add_year ( int n);                 // add n years
          void print();                           // print a date
};
```

The *public* specifier separates the class body into two parts. The names in the first, *private*, including data members **day**, **month** and **year**, part can be used only by member functions. The second, *public*, including member functions *init_date*, *add_year* and *print*, part constitutes the public interface to objects of the class. A *struct* is simply a *class* whose members are public by default; member functions can be defined and used exactly as before.

The *private* part of the members can be used only by member functions. In Example 3-2, the definition of member function is:

```
void init_date ( int d, int m, int y)
{   day = d;   month = m;   year = y;   }
```

Here, **day**, **month** and **year** are all data members of class **Date**. The function *init_date* is a member function of class **Date**. The members **day**, **month** and **year** can be accessed by function *init_date*.

However, non-member functions are barred from using private members. For example:

```
int main()
{
    Date d;
    d.year -= 200;    //error: Date::year is private
}
```

By default, all members of a class declared with the class keyword have private access for all its members. Therefore, any member that is declared before one other class specifier automatically has private access.

A main function is not a member function of class **Date**.

The *public* part of the members can be used by member functions of the class or other functions. For example,

```
int main()
{
    Date d;
    d.init_date (2, 6, 1999)    //ok:   init_date is public
}
```

Structures vs Classes

A *struct* in *C* is defined as a fixed collection of components, wherein the components can be of different types. This definition of components in a *struct* includes only member variables. However, a C++ *struct* is very similar to a *class*. By definition, a *struct* is a class in which members are by default public; that is,

struct s {······};

is simple shorthand for

class s { public: ······};

The access specifier ***private***: can be used to say that the members following are private, just as ***public***: says that the members following are public.

 A *class* resembles a *struct* with just one difference—all struct members are public by default, but all class members are private by default.

Except for the different names, the declarations shown in Figure 3-2 are equivalent.

```
struct Date{                    class Date{
    private:                        int day, month, year;
        int day, month, year;   public:
    public:                         void add_year(int n);
        void add_year(int n);   };
};
```

Figure 3-2   Declarations of struct Date and class Date

3.5 Constructors

The use of the function, such as *init_date,* to provide initialization for class objects is inelegant. Because it is nowhere stated that an object must be initialized, a programmer can forget to do so — or do so twice (often with equally disastrous results).

A better approach is to allow the programmer to declare a function with the explicit purpose of initializing objects. Since such a function constructs values of a given type, it is called a *constructor*.

A *constructor* is a member function that is implicitly invoked whenever a class object is created.

Definition of Constructors

A constructor is recognised by having the same name as the class itself. It declares for the purpose of initialising objects. For example,

//In the Date.h file, a constructor is added
class Date {
  public:
    Date(int, int, int);   *//constructor with parameters*
    void add_year(int);
    void print();
  private:
    int day, month, year;
};
//Definition of the constructor outside the class in the Date.cpp file
Date::Date(int d, int m, int y)
{
    day = d; month = m; year = y;
}

Initializing Objects with Constructors

When an object of the **Date** type is first created, we want its data to be initialized to 18, 11, 2010, for example,

Date   today (18, 11, 2010);

This statement means that the constructor of class **Date** is called automatically and data members **day**, **month** and **year** of **Date** are initialized by values 18, 11 and 2010 when object **today** is created.

When a class has a constructor, all objects of that class will be initialized. If the constructor requires arguments, these arguments must be supplied. For example,

int main()
{
    Date   today (18, 11, 2010);   *//ok*
    Date   tomorrow (19, 11, 2010);   *//ok*
    Date   my_birthday;   *//error: initializer missing*
    Date   xmax (25,12);   *//error: 3rd argument missing*
}

3.5.1   Overloading Constructors

To guarantee that the members of a class are initialized, you use constructors. There are two types of constructors: with parameters and without parameters. The constructor without parameters is called the ***default constructor***.

It is often nice to provide several ways of initializing a class object. This can be done by providing several constructors. The constructors obey the same overloading rules as do other functions. As long as the constructors sufficiently differ in their parameter types, the compiler can select the correct one for each use.

For example,

```
//----------------------------------------------------------------------
// This program overloads the constructors of class Date.
//----------------------------------------------------------------------
1   class Date{
2   public:
3       Date(int, int, int);        //constructor with three integer parameters
4       Date(int, int);             // constructor with two integer parameters
5       Date(int);                  // constructor with one integer parameter
6       Date(const *char);          // constructor with a char pointer parameter
7       //....
8   private:
9       int day, month, year;
10  };
11
12  int main()
13  {
14      Date today(18,11, 2010);
15          Date xmas(25, 12);
16          Date d ("18-11-2010");
17  }
```

3.5.2 Constructors with Default Parameters

A constructor can also have default parameters. In such class, the rules for the declaring formal parameters are the same as those of for declaring default parameters in a function. Moreover, actual parameters to a constructor with default parameters are passed according to the rules for functions with default parameters.

> A constructor that has no parameter, or has all default parameters, is called the ***default constructor.***

Example 3-4: An example of the **Date** class with default constructor.

```
//----------------------------------------------------------------------
// File: Date.h (specification file)
// This program defines the Date class.
//----------------------------------------------------------------------
1   class Date {
2   public:
3       //constructor with default parameters
4       Date(int = 1, int = 1, int = 2000);
5       void add_year (int);
6       void print();
7
```

```
8    private:
9         int day;
10        int month;
11        int year;
12   };
```

In the implementation file, the definition of the constructor is the same as the definition of the constructors with parameters.

//--
// File: **Date.cpp** (implementation file)
// This program defines the implementation detail of the Date class.
//--

```
1    #include  "Date.h"           //referring the declaration of the class
2    #include <iostream>
3    using namespace std;
4
5    //Definition of a constructor
6    Date::Date ( int d, int m, int y)
7    {
8         day = d;   month = m;   year = y;
9    }
10   // Add a number to member year
11   void Date::add_year ( int n)
12   {
13        year += n;
14   }
15   //Print a date in the English standard format
16   void Date::print ()
17   {
18        cout<< month <<"-"<<day << "-" <<year;
19   }
```

In the user file, you can declare **Date** objects with zero, one, two, or three arguments, as follows:

//--
// File: **user.cpp** (user file)
// This program creates Date objects with different parameter lists.
//--

```
1   void main()
2   {
3        Date  day;
4        Date  myday(23);
5        Date  xmas(25, 12);
```

6   Date   today(18, 11, 2010);
7   }

In Line 3, the members of *day* are initialized to : day = 1, month = 1, year = 2000. In Line 4, the members of *myday* are initialized to : day = 23, month = 1, year = 2000. In Line 5, the members of *xmas* are initialized to : day = 25, month = 12, year = 2000. In Line 6, the members of *today* are initialized to : day = 18, month = 11, year = 2010.

Sometimes we also want to define variables of type Date without initializing them, for example,

```
class Date {
public:
    Date()          // constructor without arguments
    { day = 1; month = 1; year =1; }
private:
    int day, month, year;
};
```

When the objects are created as follows,

   Date today, someday;

there is no constructor in the above statement, but our definitions worked just fine. How could they work without a constructor? Because an implicit no-argument constructor is built into the program automatically by the compiler, and it is default constructor that created the objects, even though we did not define it in the class.

If you declare an object and want the default constructor to be executed, the declaration

   Date day;   //ok: invoking constructor Date()

is legal. However, you declare the *day* object in the following format:

   Data day();   //error

it is illegal.

- A constructor is similar to a member function, but with the following differences:
- No return type.
- No return statement.
- The name of a constructor is the same as the name of the class.
- A class can have more than one constructor. However, all constructors of a class have the same name.
- If a class has more than one constructor, the constructor must have different formal parameter lists. That is, either they have a different number of the formal parameters or, if the number of formal parameters is the same, then the data type of the formal parameters must differ in at least one position.
- Constructors execute automatically when a class object is created.
- Which constructor executes depends on the types of values passed to the class object when the class object is declared.

3.6 Destructors

3.6.1 Definition of Destructors

A constructor initializes an object. In other words, it creates the environment in which the member functions operate. Sometimes, creating that environment involves acquiring a resource, such as a file, a lock, or some memory, that must be released after use. Thus, some classes need a function that is guaranteed to be invoked when an object is destroyed in a manner similar to the way a constructor is guaranteed to be invoked when an object is created.

Destructors are usually used to deallocate memory and do other cleanup for a class object and its class members when the object is destroyed. A destructor is called for a class object when that object passes out of scope or is explicitly deleted.

> A **destructor** is a member function that is used to clean up and release resources.

A destructor is a member function with the same name as its class prefixed by a ~ (tilde). For example:

```
1    //definition a destructor of class Date
2    class Date {
3    public:
4        Date(int, int, int);      //constructor
5        //other member functions
6        ~Date();                  //destructor
7    private:
8        //data members
9    };
```

A destructor takes no arguments and has no return type. Its address cannot be taken. Destructors cannot be declared const, volatile, const volatile or static. A destructor can be declared virtual or pure virtual (see Section 7.2.4).

If no user-defined destructor exists for a class and one is needed, the compiler implicitly declares a destructor. This implicitly declared destructor is an inline public member of its class.

The compiler will implicitly define an implicitly declared destructor when the compiler uses the destructor to destroy an object of the destructor's class type. Suppose a class **Date** has an implicitly declared destructor. The following is equivalent to the function the compiler would implicitly define for **Date**:

```
Date::~Date() { }      //definition outside the Date class
```
or
```
~Date()  { }           //definition inside the Date class
```

Chapter 3 Classes and Objects (I)

 A destructor is a member function, but with the following differences:
- No return type.
- No return statement.
- No parameters
- The name of a destructor is the same as the name of the class prefixed by a ~ (tilde).
- A class can have only one destructor – destructor overloading is not allowed.
- Destructors execute automatically when a class object goes out of the scope.

UML Class Diagram for Class Date

The Unified Modelling Language (UML) is now the most widely used graphical representations scheme for modelling object-oriented system. It has indeed unified the various popular national schemes. Those who design systems use the language (in the form of diagram) to model their system, as we do throughout this book.

In the UML, each class is modelled in a class diagram as a rectangle with three compartments. Figure 3-3 presents a UML class diagram for class Date in above example. The top compartment describes the name of the Date class. The middle compartment contains the data members **day**, **month** and **year**. The bottom compartment presents the member functions *add_year* and *print*. Of course, the constructor and destructor of class **Date** are also included within this compartment. The plus sign (+) in front of member names indicates that these members are public. The minus sign (-) indicates private members.

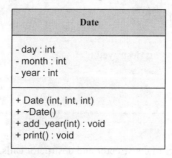

Figure 3-3 UML class diagram for class Date

3.6.2 Order of Constructor and Destructor Calls

Constructors and destructor are called automatically by compiler. The order in which these function calls are made depends on the order in which execution enters and leaves the scopes in which objects are instantiated. Generally, the destructor calls are made in the reverse order of the corresponding constructor calls.

Example 3-5: Order of constructor and destructor calls.

//--
// File: **ObjCallOrder.h**
// This program presents the specification of class ObjCallOrder.

```
1    class ObjCallOrder {
2    public:
3         ObjCallOrder(int);        //constructor
4         ~ObjCallOrder();          //destructor
5    private:
6         int data;
7    };
```

// File: **ObjCallOrder.cpp**
// This program presents the implementation of class ObjCallOrder.

```
1    #include "ObjCallOrder.h"
2    #include <iostream>
3    using namespace std;
4
5    ObjCallOrder::ObjCallOrder(int d)
6    {
7         data = d;
8         cout <<"Object "<<data<<" is constructed\n";
9    }
10
11   ObjCallOrder::~ObjCallOrder()
12   {
13        cout<<"Object "<<data<<" is destroyed\n";
14   }
```

// File: **example3_5.cpp**
// This program demonstrates the order of constructor and destructor calls.

```
1    #include "ObjCallOrder.h"
2    #include <iostream>
3    using namespace std;
4
5    // a sub-function using the class object
6    void foo()
7    {
8         ObjCallOrder furthObj(200);    //local object
9    }
10
11   ObjCallOrder firstObj(1);            //global object
```

```
12
13    int main()
14    {
15        ObjCallOrder secondObj(100);
16
17        for (int i = 2; i <= 4; i++)
18        {
19            ObjCallOrder loopObj(i);    //local objects
20        }
21
22        foo();
23
24        return 0;
25    }
26
```

Result:

1   Object 1 is constructed
2   Object 100 is constructed
3   Object 2 is constructed
4   Object 2 is destroyed
5   Object 3 is constructed
6   Object 3 is destroyed
7   Object 4 is constructed
8   Object 4 is destroyed
9   Object 200 is constructed
10 Object 200 is destroyed
11 Object 100 is destroyed
12 Object 1 is destroyed

The above program demonstrates the order in which constructors and destructors are called for objects of class **ObjCallOrder** in several scopes.

The program defines the object **firstObj** in Line 11 (example 3-5.cpp) in global scope. Its constructor is called as the program begins execution and its destructor is called at program termination after all other objects are destroyed.

The main function declares four local objects. One is **secondObj** in Line 15. The other three objects **loopObj** in Line 19 are in the *for* loop body. The constructor for **secondObj** is called when execution reaches the point where this object is declared. The destructor for **secondObj** is called when the end of **main** is reached. The constructors for **loopObj** are called when execution enters the loop body and the destructors for **loopObj** are called when execution leaves the loop body.

The **foo** function declares object **furthObj**. The constructor for **furthObj** is called when

Object-Oriented Programming in C++

execution reaches the point where **furthOjb** is declared. The destructor for **furthObj** is called when the end of **foo** is reached.

Think These Over
1. What is the purpose of using constructors and destructors in a class?
2. What are the differences between constructors/destructors and member functions of the class?

3.7 Encapsulation

Encapsulation is one of the fundamental principles of object-oriented programming. It is the grouping of related ideas into one unit, which can thereafter be referred to by a single name. The process is combining data and functions into a single unit called **class**.

Encapsulation means that all of the object's data is contained and hidden in the object and access to it restricted to members of that class, that is, the programmer cannot directly access the data. Data is only accessible through the functions existing inside the class.

Example 3-6: Design the **Counter** objects by using the concept of encapsulation.

//--
// File: **Counter.h**
// The program presents the declaration of class Counter.
//--
```
1   class Counter{
2   public:
3       // constructor with default argument
4       Counter(int i = 0);
5       // interface to outside world
6       void add();
7       // interface to outside world
8       int getSum();
9   private:
10      // hidden data from outside world
11      int sum;
12  };
```
//--
// File: **Counter.cpp**
// The program presents the implementation of class Counter and tests it.
//--
```
1   #include "Counter.h"
2   #include <iostream>
3   using namespace std;
4
```

```
5   Counter::Counter(int i)
6   { sum = i;   }
7   void Counter::add()
8   { sum++;      }
9   int Counter::getSum()
10  { return sum;}
11
12  int main()
13  {
14      Counter c1;
15      c1.add();
16      c1.add();
17      c1.add();
18      cout << "The sum of the counter is " << c1.getSum() <<endl;
19
20      Counter c2(20);
21      c2.add();
22      c2.add();
23      cout << "The sum of the counter is " << c2.getSum() <<endl;
24
25      return 0;
26  }
```

Result:

1   The sum of the counter is 3
2   The sum of the counter is 22

The **Counter** class increments a number by one successively, and returns the sum. The public member functions **add** and **getSum** are the interfaces to the outside world and a user needs to know them to use the class. The private member **sum** is something that is hidden from the outside world, but is needed for the class to operate properly.

In a word, Data encapsulation led to the important concept of data hiding. Data hiding is the implementation details of a class that are hidden from the user.

3.8 Case Study: A GradeBook Class

Now, let us take an example to demonstrate how to create a class and how to use it.

We design a grade book that helps a teacher to efficiently manage his/her student grades. A class, named GradeBook, has the following requirements:

- Store the attributes of a course, i.e. course name, teacher and hour.
- Display these attributes.
- Modify the attributes.

According to the above statements, the UML diagram of the GradeBook class is shown in Figure 3-4.

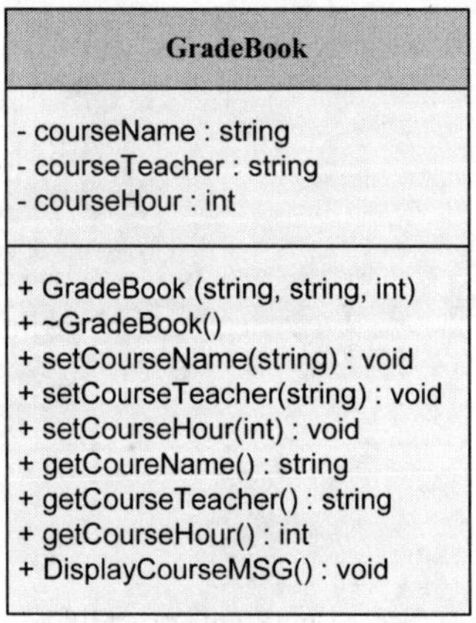

```
┌─────────────────────────────────────────┐
│              GradeBook                  │
├─────────────────────────────────────────┤
│ - courseName : string                   │
│ - courseTeacher : string                │
│ - courseHour : int                      │
├─────────────────────────────────────────┤
│ + GradeBook (string, string, int)       │
│ + ~GradeBook()                          │
│ + setCourseName(string) : void          │
│ + setCourseTeacher(string) : void       │
│ + setCourseHour(int) : void             │
│ + getCoureName() : string               │
│ + getCourseTeacher() : string           │
│ + getCourseHour() : int                 │
│ + DisplayCourseMSG() : void             │
└─────────────────────────────────────────┘
```

Figure 3-4   The UML diagram for the GradeBook class

Example 3-7. Demonstrate how to define a GradeBook class and use it.

//--
// File: **GradeBook.h**
// The program presents the declaration of class GradeBook.
//--

```
1   class GradeBook {
2   public:
3       // constructor initializes data members courseName, courseTeacher and courseHour
4       GradeBook(string coursename, string courseteacher, int coursehour );
5       // three set functions to reset the values of courseName, courseTeacher and courseHour
6       void setCourseName( string name );
7       void setCourseTeacher( string teachername );
8       void setCourseHour( int classhour);
9       // three get function to obtain the values of courseName, courseTeacher and courseHour
10      string getCourseName();
11      string getCourseTeacher();
12      int getCourseHour();
13      // display course messages and to the GradeBook user
14      void displayCouseMSG();
15      ~GradeBook();
16  private:
17      string courseName;         // course name
18      string courseTeacher;      // course teacher
19      int courseHour;            // course hour
20  };
```

First, we declare three private data members **courseName**, **courseTeacher** and **courseHour** in the GradeBook. These data members represent the attributes of the class. Then, three *set* member functions are declared to reset the attribute values, and three *get* functions are used to obtain the attribute values. Finally, a *displayCourseMSG* function is declared.

//--
// File: **GradeBook.cpp**
// The program presents the implementation of class GradeBook and test it.
//--

```
1  #include <iostream>
2  using namespace std;
3  #include "GradeBook.h"     // include definition of class GradeBook
4  GradeBook::GradeBook(string coursename, string courseteacher, int coursehour )
5  {
6      setCourseName( coursename );       // call set function to initialize courseName
7      setTeacherName(courseteacher); // call set function to initialize courseTeacher
8      setClassHour( coursehour);         // call set function to initialize courseHour
9  }
10
11 void GradeBook::setCourseName( string name )
12 {
13    courseName = name; // store the course name in the object
14 }
15 void GradeBook::setCourseTeacher( string teacher )
16 {
17    courseTeacher = teacher; // store the course teacher in the object
18 }
19 void setCourseHour( int hour)
20 {
21    courseHour = hour;   // store the course teacher in the object
22 }
23
24 string getCourseName()
25 {
26    return courseName; // return object's courseName
27 }
28 string getCourseTeacher()
29 {
30    return courseTeacher; // return object's courseTeacher
31 }
32 int getCourseHour()
33 {
```

```
34    return courseHour; // return object's courseHour
35 }
36
37 void displayMessage()
38 {
39    cout << "Welcome to the grade book for\n" << getCourseName() << "!" << endl;
40    cout << "Teacher: " << getCourseTeacher() << "; " << "Hour: " << getCourseHour() <<endl;
41 }
42
43 GradeBook::~GradeBook()
44 {   cout << "Destroying an object\n";   }
45 int main()
46 {
47    // create two GradeBook objects
48    GradeBook gradeBook1( "Object-Oriented Programming in C++", "Christina", 48);
49    gradeBook1.displayCourseMSG();
50
51    GradeBook gradeBook2( "JAVA programming", "Ellen", 40 );
52    gradeBook2.displayCourseMSG();
53    return 0;
54 }
```

Result:

1 Welcome to the grade book for Object-Oriented Programming in C++
2 Teacher: Christina; Hour: 48
3 Welcome to the grade book for JAVA programming
4 Teacher: Ellen; Hour: 40
5 Destroying an object
6 Destroying an object

In the GradeBook.cpp file, we present the definitions of member functions outside the class. To test this class, we create two objects of the class, namely gradeBook1 and gradeBook2, in the *main* function. The two objects access class *displayCourseMSG* member function to display the information of these two courses respectively.

Word Tips

access   *vt*.   存取，访问   bar   *vt*.   阻止
aggregate   *vt./vi*. 聚合   call   *vt./vi*. 调用
allocate   *vt./vi*   分配   compiler   *n*.   编译器
arrow   *n*.   箭头   component   *n*.   组成部分
assume   *vt*. 假定   conflict   *vi*.   冲突

constitute	*vt.* 组成，构成	memory	*n.*	记忆内存
construct	*vt.* 构造	modify	*vt./vi.*	修改
constructor	*n.* 构造函数	operator	*n.*	运算符
debug	*vt.* 排除故障	overload	*vt.*	重载
declaration	*n.* 声明	parameter	*n.*	参数
declare	*vt./vi.* 声明	pointer	*n.*	指针
default	*n.* 缺省，默认	prefix	*n.*	前缀
definition	*n.* 定义	previous	*adj.*	先前的
delimit	*vt* 定界	prone	*adj.*	有可能的
disastrous	*adj.* 灾难性的	recognize	*vt./vi.*	认为
equivalent	*adj.* 等价的	release	*vt.*	释放
execution	*n.* 执行	representation	*n.*	代表
explicit	*adj.* 明显的	reserve	*vt.*	预留
format	*n.* 格式	respectively	*adv.*	分别地，各自地
function	*n.* 函数	scope	*n.*	作用域
guarantee	*vt.* 保证	semicolon	*n.*	分号
inelegant	*adj.* 粗糙的	specify	*vt.*	详述，指定
initialize	*vt.* 初始化	structure	*n.*	结构体
invoke	*vt.* 调用	tag	*n.*	标志
loop	*n.* 循环	value	*n.*	值
manipulation	*n.* 操作	variable	*n.*	变量

Exercises

1. Mark the following statements as true or false and give reasons.

(1) Objects normally are not allowed to know how other objects are implemented.

(2) Class is also referred to as programmer-defined types.

(3) A member function defined outside of the class where it is declared does not have class scope.

(4) A member function with the same name as the class and preceded with a ~ is called the constructor function.

(5) Member function cannot overloaded.

(6) Operators "." and " ->" are used to accessed class members.

(7) Header files provide the complete definition of a class's implementation details.

(8) Each class definition is normally placed in a header file.

(9) The default access mode for members of a class is private.

(10) The labels public or private can be repeated in a class definition.

(11) Member functions labelled public can only be accessed from outside the class.

(12) Access to members of a struct can be defined public or private.

(13) Like any function, constructors can return a value.

(14) Constructors cannot be overloaded.

(15) A constructor is a member function with the same names as the class.
(16) A class has only one constructor.
(17) There can only be one default constructor per class.
(18) Constructor can contain default arguments in the parameter list.
(19) A destructor is called when an object is removed from memory.
(20) Like a constructor, destructor can take arguments.
(21) A destructor may return a value.
(22) The destructor can be overloaded.
(23) Constructors and destructors are called automatically.
(24) The programmer must always provide a constructor and destructor for a class.

2. Find the syntax errors in the definitions of the following classes.

(1)
```
class AA {
    public:
        void AA(int, int);
        int sum();
    private:
        int x = 0;
        int y;
};
```

(2)
```
class BB   {
    public:
        BB(int, int);
        print();
    private:
        int x, y;
}
```

(3)
```
class CC   {
    public:
        CC();
        CC(int, int);
    private:
        Int x, y;
};
int main()
{   CC c1(4);
    CC c2(4, 5);
    CC c3;
    cout<< c2.x<<c2.y<<endl;
    return 0; }
```

3. Consider the following declarations:
```
class Point {
    public:
```

```
                Point();
                Point(int, int);
                void move(int, int);
                void print();
                ~Point();
    private:
                int x, y;
};
```
and assume that the following statements are in a user program:

```
                Point point1;
                Point point2(10, 20);
```

(1) How many members does class **Point** have?

(2) How many *private* members does class **Point** have?

(3) How many constructors does class **Point** have?

(4) Write the definition of default constructor of class **Point** so that the *private* member variables are initializing by 0.

(5) Write the definition of the constructor with arguments **xx** and **yy** so that the *private* member variables are initializing by the values of xx and **yy**.

(6) Write the definition of member function *move* with arguments **newX** and **newY** so that **x** is reset to **newX** and **y** is reset to **newY**.

(7) Write the definition of member function *print* that output the values of **x** and **y**.

(8) Write a *main* function to test class **Point**.

4. Write the definition of a *Word* class that implements the functions of adding a meaning, getting a meaning, getting the number of word meanings and outputting all information of a word.

5. Write a **TimeDemo** class that includes the following properties:

(1) three data members of **hour**, **minute** and **second**

(2) three functions set to set three data members respectively

(3) time output in form of 12hours or 24hours, such as

    12hours: 9:45AM   3:10PM
    24hours: 9:45   15:10

(4) a constructor with default parameters(default time is 0:0:0)

6. Define a **Rectangle** class with data members **length** and **width**, each of which defaults to 1.

(1) two public member functions calculate the perimeter and area of the rectangle

(2) public set functions change the values of **length** and **width** and verify that **length** and **width** are each integer larger than 0 and less than 50

(3) public get functions return the values of **length** and **width**

Chapter 4
Classes and Objects (II)
— *Further Definition of Class Members and Objects* —

> *Those types are not "abstract",*
> *they are as real as* int *and* float.
> —*Doug Mellroy*

Objectives

- To specify *const* members and *const* objects
- To be able to use *static* members
- To understand the use of the *this* pointer
- To be able to create objects of other classes as class members
- To understand the purpose of *friend* functions and *friend* classes
- To understand how to use shallow copy and deep copy

4.1 Constant Member Functions and Constant Objects

The **Date** defined so far provides member functions for giving a value and changing it. Unfortunately, we did not provide a way of examining the value of a Date. This problem can easily be remedied by adding functions for reading the day, month, and year.

Example 4-1: Definition of a Date class with constant member functions.

```
//-------------------------------------------------------------------------
// File: Date.h (specification file)
// This program defines a Date class with constant member functions.
//-------------------------------------------------------------------------
1    class Date {
2    public:
3        Date(int = 1, int = 1, int = 2000);   //constructor with default parameters
4        void add_year (int);
5        int getDay()   const { return day; }
6        int getMonth() const   { return month; }
7        int getYear()  const { return year;   }
```

8   private:
9   int day;
10   int month;
11   int year;
12   };

In Lines 5-7, the member functions are specified as const in their prototypes. This means that the compiler does not allow these functions to modify data members of an object.

//---
// File: **Date.cpp** (Implementation file)
// This program implements the details of a Date class and test it.
//---

```
1  #include <iostream>
2  using namespace std;
3  #include "Date.h"
4
5  Date::Date(int d, int m, int y)
6  { day = d; month = m; year = y;}
7  void Date::add_year(int n)
8  { year += n; }
9
10 int Date::getDay() const
11 {   return day;   }
12 int Date::getMonth() const
13 {   return month;   }
14 int Date::getYear() const
15 {   return year;   }
16
17 int main()
18 {
19    Date today;
20    today.add_year (15);
21    cout << today.getDay() << " / " << today.getMonth() << " / " << today.getYear() << endl;
22    return 0;
23 }
```

Result:

1 / 1 / 2015

The *const* keyword is placed after the (empty) parameter list in the function declarations. It indicates that these functions do not modify the state of a **Date**.

Naturally, the compiler will catch accidental attempts to violate this promise. If the statement in Line 7 is as follows,

```
int getYear()    const
{
    return year++;    //error:attempt to change member value in const function
}
```

the compiler will tell you an error information.

When a *const* member function is defined outside its class, the *const* suffix is required. In other words, the const is part of the type of *Date::getDay()*.

```
int Date::getYear()    const
{
    return    year;    //ok
}
```

Some objects need to be modifiable and some do not. The *const* keyword can be used to specify that an object is not modifiable and that any attempt to modify the object should result in a compilation error. A *const* member function can be invoked for both const and non-const objects, whereas a non-const member function can be invoked only for non-const objects. For example,

```
int main()
{
    Date d1;                        // d1 is a non-const object
    const Date& d2 = d1;            // d2 is a const object
    int i = d1.getYear();           //ok
    d1.add_year(2);                 //ok
    int j = d2.getYear();           //ok
    d2.add_year(2);                 //error: cannot change value of const d2
    return 0;
}
```

4.2 *this* Pointers

An object's member function can manipulate the object's data. How do member functions know which object's data members to manipulate? Now, let us have a look at the following program that tests the Date class defined in Example 4-1.

```
1    int main()
2    {
3        Date today;
4        cout << "The size of Today object:" << sizeof(today) << " Its address " << &today << endl;
5        cout << today.getDay() << " / " << today.getMonth() << " / " << today.getYear() << endl;
6        Date tomorrow(5, 4);
```

```
7    cout << "The size of Tomorrow object:" << sizeof(tomorrow) << " Its address "
8                   << &tomorrow <<endl;
9    cout << tomorrow.getDay() << "/" << tomorrow.getMonth() << "/" << tomorrow.getYear() <<endl;
10   return 0;
11 }
```

Result:

1 The size of Today object:12 Its address 0012FF74
2 1 / 1 / 2000
3 The size of Tomorrow object:12 Its address 0012FF68
4 5/4/2000

Any Object has access to its own address through a pointer called *this*. But the this pointer of an object is not part of the objects itself – i.e. the size of the memory taken by the *this* pointer is not reflect in the result of a **sizeof** operation on the objects. For example, both the size of the **today** and **tomorrow** objects are 12 in this example.

Rather, the *this* pointer is passed as an implicit argument to each of the non-static member functions of a **class**, **struct**, or **union** type, when a non-static member function is called for an object. Instead, static member functions do not have a *this* pointer. For example, the following function call

   today.add_year(15) ;

can be interpreted this way:

   add_year(&today, 15);

The object's address is available from within the member function as the *this* pointer. Most uses of *this* are implicit. It is legal, though unnecessary, to explicitly use *this* when referring to members of the class. For example,

```
void Date::add_year (int n)
   {
       year    += n;        // ok
       this->year   += n;       //ok
       (*this).year   += n;     //ok
   }
```

These three statements are equivalent. The expression *this is commonly used to return the current object. For example,

```
1  #include <iostream>
2  using namespace std;
3  class Date{
4  public:
5     Date();
6     Date& setDate(int, int, int);
7     int getYear() const;
```

```
 8    private:
 9       int day, month, year;
10   };
11   Date::Date()
12   { day = 1; month = 1; year = 2015;}
13   Date& Date::setDate(int d, int m, int y)
14   {
15      day = d; month = m; year = y;
16      return *this;
17   }
18   int Date::getYear() const
19   { return this->year;   }
20   int main()
21   {
22      Date today;
23      cout << today.setDate(5, 5, 2015).getYear() <<endl;
24      return 0;
25   }
```

Result:

1   2015

You may notice that the *setDate* function that returns a reference to a **Date** object. Line 16 return the current object from member function *setDate* by using ***this**. Using the ***this*** pointer enables cascaded function calls in which multiple functions are invoked. In Line 23, the **today** object first invokes the *setDate* member function. Then, the **today** object with new data invokes *getYear* function.

 The ***this*** pointer is not an ordinary variable. Because the ***this*** pointer is non-modifiable, assignments to **this** are not allowed.

4.3 Static Members

Class members can be declared using the storage class specifier *static* in the class member list. Only one copy of the static member is shared by all objects of a class in a program. When you declare an object of a class having a static member, the static member is not part of the class object.

Static member is a variable that is part of a class, yet is not part of an object of that class.

A typical use of static members is for recording data common to all objects of a class. For example, you can use a static data member as a counter to store the number of objects of a particular class type that are created. Each time a new object is created, this static data member can be incremented to keep track of the total number of objects.

Example 4-2_1: Definition of a **student** class without static members.

//--
// File: **student.h** (specification file)
// This program describes the definition of a **student** class.
//--

```
1   class student{
2   public:
3       student();
4       void print() const;
5   private:
6       int count;              // a counter recording the number of objects
7       int studentNo;          // an identifier of a student
8   };
```

//--
// File: **student.cpp** (implementation file)
// The program presents the implementation detail of the **student** class
//--

```
1   #include "student.h"
2   #include <iostream>
3   using namespace std;
4
5   student::student()
6   {
7       count = 0;
8       studentNo = count;
9   }
10
11  void student::print() const
12  { cout<<"student = "<<studentNo<<" count ="<<count<<endl;}
```

//--
// File: **example4_2_1.cpp**
// This program creates student objects and accesses their members.
//--

```
1   #include "student.h"
2   #include <iostream>
3   using namespace std;
4   void main()
5   {
```

```
6         student st1;
7         st1.print();
8         cout<<"**********\n";
9         student st2;
10        st1.print();
11        st2.print();
12        cout<<"**********\n";
13        student st3;
14        st1.print();
15        st2.print();
16        st3.print();
17    }
```

Result:

```
1     student=0 count=0
2     **********
3     student=0 count=0
4     student=0 count=0
5     **********
6     student=0 count=0
7     student=0 count=0
8     student=0 count=0
```

When we instantiate a class object, each object gets its own copy of all normal data members. In this case, because we have declared three **student** class objects, we end up with three copies of **count** — one inside **st1**, the second inside **st2**, and the third inside **st3**. They should have different values because the **count** variable is used to keep track of the total number of objects. However, they get the same value 0.

4.3.1 Static Data Members

To keep track of the number of objects, one way to think about it is that all objects of a class share the static variables.

Example 4-2_2: Definition of a **student** class with static members.

```
//----------------------------------------------------------------------
// File: student.h (specification file)
// This program describes the definition of a student class.
//----------------------------------------------------------------------
1     class student{
2     public:
3         student();
4         void print() const;
5         static int getCount();        //a static member function
```

```
6    private:
7        static int count;        // a static data member
8        int studentNo;           // an identifier of a student
9    };
```

//--
// File: **student.cpp** (implementation file)
// The program presents the implementation detail of the **student** class
//--

```
1    #include "student.h"
2    #include <iostream>
3    using namespace std;
4
5    int student::count = 0;       //definition outside class declaration
6
7    student::student()
8    {
9        count ++;                 //instead of count=0
10       studentNo = count;
11   }
12
13   void student::print() const
14   {
15       cout<<"student = "<<studentNo<<" count = "<<count<<endl;
16   }
17
18   int student::getCount()
19   {
20       return count;
21   }
```

//--
// File: **example4_2_2.cpp**
// This program creates student objects and obtains the number of student objects.
//--

```
1    #include "student.h"
2    #include <iostream>
3    using namespace std;
4
5    int main()
6    {
7        student st1;
8        st1.print();
9        cout<<"***********\n";
```

Object-Oriented Programming in C++

```
10        student st2;
11        st1.print();
12        st2.print();
13        cout<<"**********\n";
14        student st3;
15        st1.print();
16        st2.print();
17        st3.print();
18        cout<<"**********\n";
19        cout<<"Student's number is "<<student::getCount()<<endl;
20    }
```

Result:

```
1     student=1 count=1
2     **********
3     student=1 count=2
4     student=2 count=2
5     **********
6     student=1 count=3
7     student=2 count=3
8     student=3 count=3
9     **********
10    Student's number is 3
```

Since the **count** variable is a static data member, **count** is shared between all objects of the class. Consequently, **st1.count** is the same as **st2.count** after st2 is declared in Line 10 of the example4_2_2.cpp.

Although you can access static members through objects of the class type, this is somewhat misleading. **st1.count** implies that **count** belongs to **st1**, and this is really not the case. **count** does not belong to any object. In fact, **count** exists even if there are no objects of the class have been instantiated!

Initializing Static Data Members

Since static data members are not part of the individual objects, you must explicitly define the static member if you want to initialize it to a non-zero value. The following statement in Example 4-2_2 initializes the static member to 0:

      int student::count = 0;

This initializer should be placed in the implementation file for the class (e.g. student.cpp of Example 4-2_2).

The declaration of a static data member in the member list of a class is not a definition. You must define the static members outside of the class declaration, in namespace scope, even if the static members are private.

90

4.3.2 Static Member Functions

While we create a normal public member function to access the **count** variable in Example 4-2_2, we then need to instantiate an object of the class type to use the function. Like static data members, static member functions are not attached to any particular object. Here is the above example with a static member function:

    static int getCount();
    {   return count;   }

Static member functions can be called directly by using the class name and the scope operator, because they are not attached to a particular object. Like static data members, they can also be called through objects of the class type, though this is not recommended. In Example 4-2_2, the static member function is invoked as follows:

```
1    void main() {
2        student st1;
3        st1.print();
4        cout<<"***********\n";
5        student st2;
6        st1.print();
7        st2.print();
8        cout<<"***********\n";
9        student st3;
10       st1.print();
11       st2.print();
12       st3.print();
13       cout<<"***********\n";
14       cout<<"Student's number is"<<student::getCount()<<endl;  }
```

Static member functions have two interesting features worth noting. First, since static member functions are not attached to any object, they have no *this* pointer! This makes sense when you think about it — the *this* pointer always points to the object that the member function is working on. Static member functions do not work on an object, so the *this* pointer is not needed.

Second, static member functions can only access static member variables. They cannot access non-static data members. This is because non-static data member must belong to a class object, and static member functions have no class object to work with!

Static member functions
- have no *this* pointer.
- cannot access non-static data members.
- do not need any object to be invoked.

Think These Over
1. Why do static data members of a class differ from the ordinary data members?
2. How do we obtain static data members?

4.4 Free Store

C++ has several distinct memory areas where objects and non-object values may be stored, and each area has different characteristics shown in Table 4-1.

Table 4-1 Characteristics of different memory areas

Memory Area	Characteristics and Object Lifetime
Const data	The const data area stores string literals and other data whose values are known at compile time. No objects of class type can exist in this area. All data in this area is available during the entire lifetime of the program. Further, all of this data is read-only, and the results of trying to modify it are undefined.
Stack	The stack stores automatic variables. Typically, allocation is much faster than for dynamic storage (heap or free store) because a memory allocation involves only pointer increment rather than more complex management. Objects are constructed immediately after memory is allocated and destroyed immediately before memory is deallocated, so there is no opportunity for programmers to directly manipulate allocated but uninitialized stack space.
Free store	The free store is one of the two dynamic memory areas, allocated/freed by new/delete. Object lifetime can be less than the time the storage is allocated; that is, free store objects can have memory allocated without being immediately initialized, and can be destroyed without the memory being immediately deallocated. During the period when the storage is allocated but outside the object's lifetime, the storage may be accessed and manipulated through a void* but none of the proto-object's non-static members or member functions may be accessed, have their addresses taken, or be otherwise manipulated.
Heap	The heap is the other dynamic memory area, allocated/freed by malloc/free and their variants. Note that while the default global new and delete might be implemented in terms of malloc and free by a particular compiler, the heap is not the same as free store and memory allocated in one area cannot be safely deallocated in the other. Memory allocated from the heap can be used for objects of class type by placement-new construction and explicit destruction. If so used, the notes about free store object lifetime apply similarly here.

Chapter 4 Classes and Objects (II)

(continued)

Memory Area	Characteristics and Object Lifetime
Global / Static	Global or static variables and objects have their storage allocated at program startup, but may not be initialized until after the program has begun executing. For instance, a static variable in a function is initialized only the first time program execution passes through its definition. The order of initialization of global variables across translation units is not defined, and special care is needed to manage dependencies between global objects (including class statics). As always, uninitialized proto-objects' storage may be accessed and manipulated through a void* but no non-static members or member functions may be used or referenced outside the object's actual lifetime.

Free store is a pool of memory available for you to allocate (and deallocate) storage for objects during the execution of your program. The new and delete operators are used to allocate and deallocate free store, respectively.

The return value from **new** is a memory address. It must be assigned to a pointer. To create an *integer* on the free store, you might write

> int *ptr = new int;
> int *ptr1 = new int(8);
> char *p = new char[10];

You can, of course, initialize the pointer at its creation with

> int *ptr;
> ptr = new int;

In either case, the **ptr** variable now points to an *integer* on the free store. You can use this like any other pointer to a variable and assign a value into that area of memory by writing

> *ptr =20;

This means, "put 20 at the value in **ptr**," or "assign the value 20 to the area on the free store to which **ptr** points".

If using the **new** keyword cannot create memory on the free store (memory is, after all, a limited resource), it returns the null pointer. You must check your pointer for null each time you request new memory.

Each time you allocate memory using the **new** keyword, you must check to make sure the pointer is not null.

When you are finished with your area of memory, you must call the ***delete*** statement on the pointer. The ***delete*** statement returns the memory to the free store. Remember that the pointer itself — as opposed to the memory to which it points—is a local variable. When the function in which it is declared returns, that pointer goes out of scope and is lost. The memory allocated by using *new* is not freed automatically, however. That memory becomes unavailable—a situation called a *memory leak*. It's called a memory leak because that memory cannot be recovered until the program ends. It is as though the memory has leaked out of your computer.

To restore the memory to the free store, you use the **delete** keyword. For example,

    delete ptr;
    delete ptr1;
    delete []p;

When you delete the pointer, what you are really doing is freeing up the memory whose address is stored in the pointer.

Objects by Using *new* and *delete* on Free Store

Just as you can create a pointer to an integer, you can create a pointer to any object. If you have declared an object of class **Date**, you can declare a pointer to that class and instantiate a Date object on the free store, just as you can make one on the stack.

Example 4-3: Date objects by using *new* and *delete*.

```
//---------------------------------------------------------------
// File: Date.h
// This program describes the definition of a Date class.
//---------------------------------------------------------------
1    class Date {
2    public:
3       Date();                    //default constructor
4       Date(int, int, int);       //constructor
5       ~Date();                   //destructor
6       int getDay() const;
7       int getMonth() const;
8       int getYear() const;
9    private:
10      int day, month, year;
11   };
//---------------------------------------------------------------
// File: Date.cpp
// The program presents the implementation detail of the Date class
//---------------------------------------------------------------
1    #include "Date.h"
2    #include <iostream>
3    using namespace std;
4
5    Date::Date()
6    {
7       day = 1; month = 3; year = 2015;
8       cout<<"Default constructor...\n";
9    }
```

```
10
11    Date::Date(int d, int m, int y)
12    {
13        day = d; month = m; year = y;
14        cout<<"constructor...\n";
15    }
16
17    Date::~Date()
18    {   cout<<"Destructor...\n";   }
19
20    int Date::getDay() const
21    {   return day;        }
22    int Date::getMonth() const
23    {   return month;      }
24    int Date::getYear() const
25    {   return year;       }
```
//--
// File: **example4_3.cpp**
// This program creates Date objects by using *new* on the free store and accesses their members.
//--
```
1     #include "Date.h"
2     #include <iostream>
3     using namespace std;
4
5     int main()
6     {
7         Date day;
8         Date today(18, 5, 2015);
9         Date *dPtr;
10        dPtr = &day;
11
12        cout<<"Constructing objects by new....\n";
13        Date *dPtr1 = new Date;
14        Date *dPtr2 = new Date(20, 3, 2000);
15
16        cout << dPtr->getDay() << "/" << dPtr->getMonth() << "/" << dPtr->getYear() << endl;
17        cout << dPtr1->getDay() << "/" << dPtr1->getMonth() << "/" << dPtr1->getYear() << endl;
18        cout << dPtr2->getDay() << "/" << dPtr2->getMonth() << "/" << dPtr2->getYear() << endl;
19
20        cout<<"Destroying objects by delete....\n";
21        delete dPtr1;
```

```
22          delete dPtr2;
23
24          cout<<"Ending execution...\n";
25          return 0;
26    }
```

Result:

```
1     Default constructor...
2     constructor...
3     Constructing objects by new....
4     Default constructor...
5     constructor...
6     1/3/2015
7     1/3/2015
8     20/3/2000
9     Destroying objects by delete....
10    Destructor...
11    Destructor...
12    Ending execution...
13    Destructor...
14    Destructor...
```

In Line 7 in example4_3.cpp, the **day** object is created on the stack whilst default constructor is called. In Line 8, the **today** object is created on the stack whilst the constructor is called. In Line 13, the **dPtr1** object is created on the free store. The default constructor sets its **day**, **month** and **year** to 1, 3, 2015. In Line 14, the **dPtr2** object is created on the free store. The constructor sets its **day**, **month** and **year** to 20, 3, 2000. When **dPtr1** and **dPtr2** are deleted in Lines 21 and 22, their destructors are called. The destructor deletes each of its member pointers. Since both **dPtr1** and **dPtr2** are pointers, the arrow operator (->) is used to access the member function *print* in Lines 16, 17 and 18.

4.5 Objects as Members of Classes

Sometimes, a class has a data member which is not simple built-in data type but aggregate built-in data. This data member is defined as another class object. When a class can have objects of other classes as members, such capability is called *composition*. Composition is a special case of the *aggregation* relationship.

> **Composition** is one form of software reusability, in which a class has objects of other classes as members.
> **Aggregation** is a *has-a* relationship and represents an ownership relationship between two objects.

When an object is created, its constructor is called automatically, so we need to specify how arguments are passed to member-object constructors. Member objects are constructed in the order in which they are declared (not in the order they are listed in the constructor's member initialize list) and before their enclosing class objects are constructed.

For example, we define a **Student** class that has the properties of age, birthday and body status. The birthday is defined as the **Date** class type. The body status is defined as the **Health** class type that includes student's height and weight. The birthday and body status are user-defined types. This means that the **Student** class own two other class objects. "a student has a birthday" is an aggregation relationship between the **Student** class and the **Date** class because a birthday may be shared by other students, whereas "a student has a body status" is a composition relationship between the **Student** class and the **Health** class. The UML diagram of their class relationship is shown in Figure 4-1. In UML, a filled diamond is denoted the composition relationship between the **Student** class and the **Health** class, and an empty diamond is denoted the aggregation relationship between the **Student** class and the **Date** class.

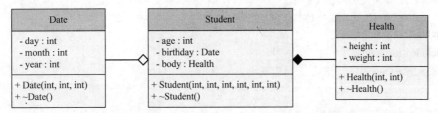

Figure 4-1   The UML diagram of the three class relationship

Example 4-4: Definition of member objects.

```
//---------------------------------------------------------------
// File: student.h
// This program describes the definition of a student class with Date and Health objects.
//---------------------------------------------------------------
1    class Date {
2    public:
3        Date(int, int, int);
4        ~Date();
5    private:
6        int day, month, year;
7    };
8
9    class Health {
10   public:
11       Health(int, int);
12       ~Health();
13   private:
14       int height;
15       int weight;
```

```
16    };
17
18    class Student  {
19       public:
20            Student(int, int, int, int, int, int);
21            ~Student();
22       private:
23            int age;
24            Date birthday;          //object of class Date
25            Health body;            //object of class Health
26    };
```

//--
// File: **student.cpp**
// The program presents the implementation detail of three classes Date, Health and Student.
//--

```
1     #include "student.h"
2     #include <iostream>
3     using namespace std;
4
5     Date::Date(int d, int m, int y)
6     {
7         day = d; month = m; year = y;
8         cout<<"Date class is created"<<endl;
9     }
10
11    Date::~Date()
12    {
13        cout<<"Date class is destroyed"<<endl;
14    }
15
16    Health::Health(int h, int w)
17    {
18        height = h; weight = w;
19        cout<<"Health class is created"<<endl;
20    }
21
22    Health::~Health()
23    {
24        cout<<"Health class is destroyed"<<endl;
25    }
26
```

```
27    Student::Student(int a, int d, int m, int y, int h, int w) :
28                    birthday(d, m, y), body(h, w)    <--------initialize ation list
29    {
30        age = a;
31        cout<<"Student class is created"<<endl;
32    }
33
34    Student::~Student()
35    {
36            cout<<"Student class is destroyed"<<endl;
37    }
```
//--
// File: **example4_4.cpp**
// This program demonstrates how two member objects are initialized.
//--
```
1    #include "student.h"
2
3    int main()
4    {
5        Student st(20, 11, 2, 1990, 180, 75);
6        return 0;
7    }
```
Result:
1 Date class is created
2 Health class is created
3 Student class is created
4 Student class is destroyed
5 Health class is destroyed
6 Date class is destroyed

The above program defines three classes **Date**, **Health** and **Student** to demonstrate objects as members of other class. Class **Student** includes private data members **age**, **birthday** and **body.** Members **birthday** and **body** are the objects of class **Date** and **Health**. The program instantiates a **Student** object and initializes. The **Student** constructor is defined as follows:

Student::Student (int a, int d, int m, int y, int h, int w) :
                    birthday(d, m, y), body(h, w)
    { age = a; }
                                → Object of the **Health** class
                                → Object of the **Date** class

The constructor takes six parameters. The colon (:) in the header separates the member initialization list (also called the member initializer) from the parameter list. The member initializer specifies the **Student** parameters being passed to the constructors of the objects. Arguments **d**, **m** and **y** are passed to the *birthday* constructor, and arguments **h** and **w** are passed to the *body* constructor.

Order of Constructors and Destructors for Objects

- **Constructors**

    The members' constructors are invoked before the body of the containing class' own constructor in executed. The constructors are invoked in the order in which the members are declared in the class rather than the order in which the members appear in the initialization list.

- **Destructors**

    When a class object containing class objects is destroyed, the body of that object's own destructor is executed first and then the members' destructors are executed in reverse order of declaration.

Object Members with Default Constructors

If a member constructor needs no arguments, the member need not be mentioned in the member initialization list. For example,

//--
// File: **example4_4_1.cpp**
// This program demonstrates the initialization of member objects with a default constructor.
//--

```
1    #include <string>
2    using namespace std;
3
4    class Score {
5    public:
6        Score():grade(0){}
7    private:
8        int grade;
9    };
10
11   class Student {
12   public:
13       Student(char* nstr);
14       ~Student();
15   private:
16       char* name;
17       Score sc;                    //object of class Score
18   };
19
20   Student::Student(char* nstr): sc()
21   {
22       name = new char[strlen(nstr) + 1];
23       strcpy(name, nstr);
```

24   }
25   Student::~Student()
26   {
27   Delete name;
28   }
29   int main()
30   {
31   Student st("wang");
32   return 0;
33   }

Members by Using Initializers

Member initializers are essential for types for which initialization differs from assignment, that is, for *member objects* of classes without default constructor, for *const* members, and for *reference* members.

For example,

//--
// File: **example4_4_2.cpp**
// This program demonstrates the initialization of special members
//--

1   class A {
2   public :
3   A(int dd, int mm, int yy, int& a) : **i(10), d(dd, mm, yy), pc(a)**{}
4   private:
5   const int i;   //*const member*
6   Date d;   //*member object*
7   int& pc;   //*reference member*
8   };
9
10 int main()
11 {
12   int i;
13   A a(23, 10, 2000, i);
14   return 0;
15 }

With annotations: initialization list → **i(10), d(dd, mm, yy), pc(a)**; pc(a) → reference variable; d(dd, mm, yy) → object of class **Date**; i(10) → const variable

To ensure that initialization requirements for the member objects are met, one of the following conditions must be met:

• The contained object's class requires no constructor.

• The contained object's class has an accessible default constructor.

• The containing class's constructors all explicitly initialize the contained object.

Object-Oriented Programming in C++

> **Think These Over**
> 1. How is an object member of the class initialized?
> 2. What is the purpose of composition?
> 3. How to draw the UML class diagram of example 4_4_1.cpp?

4.6 Copy Members

When declaring a class object, you can initialize it by using the values of an existing object of the same class.

For example, consider the following statement:

Date today (18, 11, 2010);

Date day = today;   or   Date day(today);

Object **today** has been declared. Object **day** is being declared and is also being initialized by using the value of object **today**. That is, the values of the data members of **today** are copied into the corresponding data members of **day**, as shown in Figure 4-2. The initialization is called default member-wise initialization. This default initialization is due to the constructor, called the *copy constructor*.

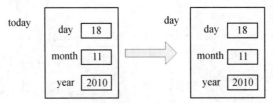

Figure 4-2   Objects **today** and **day**

Note that a **copy constructor** is called whenever a new variable is created from an existing object.

4.6.1 Definition of Copy Constructors

The *copy constructor* takes a *reference* to a *const* parameter. It is const to guarantee that the copy constructor does not change it, and it is a reference because a value parameter would require making a copy, which would invoke the copy constructor, which would make a copy of its parameter.

Example 4-5: Definition of a copy constructor.

```
//-----------------------------------------------------------------------------
// File: Date.h
// This program describes the definition of a Date class.
//-----------------------------------------------------------------------------
```

Chapter 4 Classes and Objects (II)

```
1    class Date {
2    public:
3      Date(int = 1, int = 1, int = 2000);
4      Date(const Date&);                    //copy constructor
5      ~Date();
6
7      void print() const;
8    private:
9       int day, month, year;
10   };
```
//---
// File: **Date.cpp**
// The program presents the implementation detail of the **Date** class
//---

```
1    #include "Date.h"
2    #include <iostream>
3    using namespace std;
4
5    //Definition of constructor with default arguments
6    Date::Date(int d, int m, int y) : day(d), month(m), year(y)
7    {
8           cout<<"constructing an object...\n";
9    }
10
11   //Definition of copy constructor
12   Date::Date(const Date& date)
13   {
14          day = date.day;
15          month = date.month;                     } Copy constructor
16          year = date.year;
17
18          cout<<"constructing a copy object...\n";
19   }
20
21   //Definition of destructor
22   Date::~Date()
23   {
24          cout<<"destroying an object...\n";
25   }
26
27   void Date::print() const
```

```
28      {
29              cout<<day<<"/"<<month<<"/"<<year<<endl;
30      }
```
//---

// File: **example4_5.cpp**

// This program demonstrates how the copy constructor works.

//---

```
1       #include "Date.h"
2       #include <iostream>
3       using namespace std;
4
5       Date f(Date d)
6       {
7               cout<<"This is a sub-function...\n";
8               d.print();
9               return d;
10      }
11
12      int main()
13      {
14              Date day1(7, 8, 2010);
15              Date day2 = day1;           //declaration of copy object
16              Date day3(day1);            // declaration of copy object
17              day1.print();
18              day2.print();
19
20              cout<<endl;
21              day3 = f(day2);             // copy constructor initializes formal parameter
22              cout<<endl;
23              cout<<"Ending the main function...\n";
24
25              return 0;
26      }
```

Result:

1 constructing an object...

2 constructing a copy object...

3 constructing a copy object...

4 7/8/2010

5 7/8/2010

6

7 constructing a copy object...

8 This is a sub-function...

9 7/8/2010

10 constructing a copy object...

11 destroying an object...

12 destroying an object...

13

14 Ending the main function...

15 destroying an object...

16 destroying an object...

17 destroying an object...

The above is an example of a copy constructor for the **Date** class, which does not really need one because the default copy constructor's action of copying data would work fine, but it shows how it works.

When Line 15 in example4_5.cpp declares a copy object **day2**, the copy constructor is called. From the result of Example 4-5, you find out that the copy constructor is called when the **f** function is called in Line 21 and when the **d** object is returned in the **f** function.

Therefore, the copy constructor happens in the following cases:

(1) A variable is declared which is initialized from another object, for example,

Date day1(7, 8, 2010);

Date day2 = day1;

(2) A value parameter is initialized from its corresponding argument, for example,

f(day2);

(3) An object is returned by a function, for example,

return d;

C++ calls a *copy constructor* to make a copy of an object in each of the above cases. If there is no copy constructor defined for the class, C++ uses the default copy constructor which copies each field, that is, makes a *shallow copy*.

4.6.2 Shallow Copy and Deep Copy

A *shallow copy* means that C++ copies each member of the class individually using the assignment operator (=). When classes are simple (e.g. do not contain any dynamically allocated memory), this works very well like the above example.

If the object has no pointers to dynamically allocated memory, a shallow copy is probably sufficient. Therefore, the default copy constructor, default assignment operator, and default destructor are fine, and you do not need to write your own.

Now we make a little changes for Example 4-5—remove the definition of *copy constructor* from Example 4-5. The program still works well. The changed program is illustrated in Example 4-6.

Example 4-6: Shallow copying the class members by using a default copy constructor.

//--
// File: **Date.h**
// This program describes the definition of a **Date** class.
//--

```
1   class Date {
2   public:
3       Date(int = 1, int = 1, int = 2000);
4       ~Date();
5
6       void print() const;
7   private:
8       int day, month, year;
9   };
```
//--
// File: **Date.cpp**
// The program presents the implementation detail of the **Date** class.
//--

```
1   #include "Date.h"
2   #include <iostream>
3   using namespace std;
4
5   //Definition of constructor with default arguments
6   Date::Date(int d, int m, int y):day(d), month(m), year(y)
7   {
8       cout<<"constructing an object...\n";
9   }
10
11  //Definition of destructor
12  Date::~Date()
13  {
14      cout<<"destroying an object...\n";
15  }
16
17  void Date::print() const
18  {
19      cout<<day<<"/"<<month<<"/"<<year<<endl;
20  }
```
//--
// File: **example4_6.cpp**
// This program demonstrates how the default copy constructor works.
//--

Chapter 4   Classes and Objects (II)

```
1    #include "Date.h"
2    #include <iostream>
3    using namespace std;
4
5    Date f(Date d)
6    {
7            cout<<"This is a sub-function...\n";
8            d.print();
9            return d;
10   }
11
12   int main()
13   {
14           Date day1(7, 8, 2010);
15           Date day2 = day1;         //calling a default copy constructor
16           Date day3(day1);          // calling a default copy constructor
17           day1.print();
18           day2.print();
19
20           cout<<endl;
21           day3 = f(day2);    // copy constructor initializes formal value parameter
22           cout<<endl;
23           cout<<"Ending the main function...\n";
24
25            return 0;
26   }
```

Result:

```
1    constructing an object...
2    7/8/2010
3    7/8/2010
4
5    This is a sub-function...
6    7/8/2010
7    destroying an object...
8    destroying an object...
9
10   Ending the main function...
11   destroying an object...
12   destroying an object...
13   destroying an object...
```

Although the above result is different with that of Example 4-5, the program works well. This is because the default copy constructor provided by the system is used when the copy objects are declared, e.g. in Lines 15 and 16 of file **example4_6.cpp**, or passed as a parameter in Line 5 or returning value in Line 9.

However, when designing classes that handle dynamically allocated memory, member-wise (shallow) copying can get us in a lot of trouble! This is because the standard pointer assignment operator just copies the address of the pointer — it does not allocate any memory or copy the contents being pointed to! Let's take a look at the following example.

Example 4-7: Shallow copying a pointer to member in a class.

//---
// File: **Date.h**
// This program describes the definition of a **Date** class.
//---

```
1   class Date {
2   public:
3       Date(int, int, int, char*);
4       ~Date();
5
6       void print() const;
7   private:
8       int day, month, year;
9       char *name;                //char pointer
10  };
```

//---
// File: **Date.cpp**
// The program presents the implementation detail of the **Date** class.
//---

```
1   #include "Date.h"
2   #include <iostream>
3   using namespace std;
4
5   //Definition of constructor
6   Date::Date(int d, int m, int y, char* nstr):day(d), month(m), year(y)
7   {
8       name = new char[strlen(nstr) + 1];    //allocate memory dynamically on free store
9       strcpy(name, nstr);
10
11      cout<<name <<" is constructing...\n";
12  }
13
14  //Definition of destructor
```

```
15    Date::~Date()
16    {
17        cout<<name <<" is destroying...\n";
18        delete name;                           //deallocating memory
19    }
20
21    void Date::print() const
22    {
23        cout<<day<<"/"<<month<<"/"<<year<<endl;
24    }
```
//--
// File: **example4_7.cpp**
// This program demonstrates what problems shallow copying causes.
//--
```
1     #include "Date.h"
2     #include <iostream>
3     using namespace std;
4
5     void f(Date d)
6     {
7         cout<<"This is a sub-function...\n";
8         d.print();
9     }
10
11    int main()
12    {
13        Date day1(7, 8, 2010, "day1");
14        Date day2 = day1;                      //declaration of copy object
15        Date day3(12, 5, 2000, "day3");
16
17        day1.print();
18        day2.print();
19
20        cout<<endl;
21        f(day2);
22        cout<<endl;
23        cout<<"Ending the main function..\n";
24
25        return 0;
26    }
```

109

The above program looks no problem, but it it contains an insidious problem that will cause the program to crash! When it is executed, the following error message is prompted.

Let's break down this example. Line 14 is the same as that in Example 4-6 and seems harmless enough as well, but it's actually the source of our problem caused the above error message. Why? As we mentioned above, object **day2** is declared by using existing object **day1**. This copy is executed by using default copy constructor (because we haven't defined our own), which does a shallow pointer copy for data member **day1.name**. Since a shallow pointer copy just copies the address of the pointer, the address of **day1.name** is copied to **day2.name**. As a result, are now both pointing to the same piece of memory, shown in Figure 4-3! When the end of the example4_7.cpp program is reached, the object **day2** is destroyed and the dynamically allocated memory, which **day2.name** points to, is cleaned up. When the object **day1** is destroyed, the **day1.name** points to the deleted (invalid) memory! So you see the error message. Likewise, the same error is caused when calling the sub-function **f**.

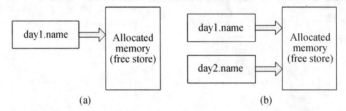

Figure 4-3   Shallow copy for *day2=day1*

(a) Date day 1(7,8,2010, "day1"); (b) Date day2=day1.

The root of this problem is the shallow copy done by the copy constructor — doing a shallow copy on pointer values in a copy constructor or overloaded assignment operator (=) is always asking for trouble. The answer to this problem is to do a deep copy on any non-null pointers being copied.

A ***deep copy*** duplicates the object or variable being pointed to so that the destination (the object being assigned to) receives its own local copy (see Figure 4-4). In this way, the

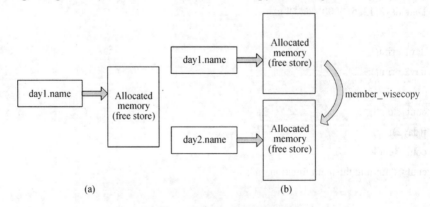

Figure 4-4   Deep copy for **day2 =day1**

(a) Date day 1(7,8,2010,"day1"); (b) Date day2=day1.

destination can do whatever it wants to its local copy and the object that was copied from will not be affected. Doing deep copies requires that we write our own copy constructors and overloaded assignment operators (see Section5.6.2).

Let's go ahead and show how this is done for the problem.

Example 4-8: Deep copying a pointer to member in a class.

//---
// File: **Date.h**
// This program describes the definition of a **Date** class.
//---

```
1   class Date {
2   public:
3       Date(int, int, int, char*);
4       Date(const Date&);        //copy constructor
5       ~Date();
6
7       void print() const;
8   private:
9       int day, month, year;
10      char *name;               //char pointer
11  };
```

//---
// File: **Date.cpp**
// The program presents the implementation detail of the **Date** class.
//---

```
1   #include "Date.h"
2   #include <iostream>
3   using namespace std;
4
5   //Definition of constructor with default arguments
6   Date::Date(int d, int m, int y, char* nstr):day(d), month(m), year(y)
7   {
8       name = new char[strlen(nstr) + 1];
9       strcpy(name, nstr);
10
11      cout<<name <<" is constructing...\n";
12  }
13
14  //Definition of copy constructor
15  Date::Date(const Date& date)
16  {
```

Object-Oriented Programming in C++

```
17      day = date.day;
18      month = date.month;
19      year = date.year;
20
21      if(date.name != NULL)
22      {
23          name = new char[strlen(date.name) + 1];     ⎫
24          strcpy(name, date.name);                    ⎬ Deep copying
25      }                                               ⎪
26      else                                            ⎪
27          name = NULL;                                ⎭
28
29      cout<<"constructing a copy object...\n";
30  }
31
32  //Definition of destructor
33  Date::~Date()
34  {
35      cout<<name <<" is destroying...\n";
36      delete name;
37  }
38
39  void Date::print() const
40  {
41      cout<<day<<"/"<<month<<"/"<<year<<endl;
42  }
```

//--
// File: **example4_8.cpp**
// This program demonstrates how deep copying works.
//--

```
1   #include "Date.h"
2   #include <iostream>
3   using namespace std;
4
5   void f(Date d)
6   {
7       cout<<"This is a sub-function...\n";
8       d.print();
9       //return d;
10  }
```

```
11
12      int main()
13      {
14          Date day1(7, 8, 2010, "day1");
15          Date day2 = day1;          //declaration of copy object
16          Date day3(12, 5, 2000, "day3");
17
18          day1.print();
19          day2.print();
20
21          cout<<endl;
22          f(day2);
23          cout<<endl;
24          cout<<"Ending the main function..\n";
25
26          return 0;
27      }
```

Result:

```
1    day1 is constructing...
2    constructing a copy object...
3    day3 is constructing...
4    7/8/2010
5    7/8/2010
6
7    constructing a copy object...
8    This is a sub-function...
9    7/8/2010
10   day1 is destroying...
11
12   Ending the main function...
13   day3 is destroying...
14   day1 is destroying...
15   day1 is destroying...
```

A class that requires deep copies generally needs:

- A *constructor* to either make an initial allocation or set the pointer to NULL.
- A *destructor* to delete the dynamically allocated memory.
- A *copy constructor* to make a copy of the dynamically allocated memory.
- An *overloaded assignment operator* to make a copy of the dynamically allocated memory.

Object-Oriented Programming in C++

Think These Over
1. What situations does a copy constructor of a class need to be defined?
2. What is the difference between shallow copy and deep copy?

4.7 Arrays of Objects

Suppose that you need to perform a number of objects. All objects are declared from the same class. A convenient way to store these objects is to use an array.

4.7.1 Initialize an Array of Objects by Using a Default Constructor

Suppose that you declare an array of 4 class objects, for example, 4 **Date** objects. you need to display of their days, months and years. You need to declare one array *arrayDate* of 4 elements each, wherein each element is an object of type **Date**.

    Date   arrayDate[4]

The above statement creates the array of four objects **arrayDate[0], arrayDate[1],…, arrayDate[3]**.

In this case, it is impractical to specify different constructors for each element. Therefore, the default constructor is used to initialize each (array) class object. If a class has constructors and you declare an object of that class, the class should have the default constructor.

In the above example, each *arrayDate* is initialized by a call of **Date::Date()**. Thus, you can use the member functions of class **Date** to display their date to each object. For example, you can write as follows:

    arrayDate[0].print();

Also you can use a loop to display the dates of four objects, such as the following:

    for (int i = 0; i < 4; i++)

        arrayDate[i].print();

The destructor for each constructed elements of an array is invoked when that array is destroyed. If the array of class objects is constructed by using *new*, then you must use *delete* explicitly to the deallocated memory. Otherwise, this leads to the problem. For example,

    Date *day = new Date[4];   //*allocate an array*

    //....

    Delete []day;   //*free an array*

Example 4-9: Initializing arrays of objects by using a default constructor.

//--
// File: **Date.h**
// This program describes the definition of a **Date** class with a default constructor.
//--
1   class Date{

Chapter 4 Classes and Objects (II)

```
2      public:
3          Date();
4          ~Date();
5          void print() const;
6      private:
7          int day, month, year;
8          static int count;
9      };
```
//---
// File: **Date.cpp**
// The program presents the implementation detail of the **Date** class.
//---
```
1    #include "Date.h"
2    #include <iostream>
3    using namespace std;
4
5    int Date::count =0;
6    Date::Date()
7    {
8        day =1; month = 1; year=2000;
9        cout<<"ArrayDate["<<count<<"] is constructed\n";
10       count++;
11   }
12
13   Date::~Date()
14   {
15       count--;
16       cout<<"ArrayDate["<<count<<"] is destroyed\n";
17
18   }
19   void Date::print() const
20   { cout<<day<<"/"<<month<<"/"<<year<<endl; }
```
//---
// File: **example4_9.cpp**
// This program demonstrates how to declare object arrays and access the element of the array.
//---
```
1    #include "Date.h"
2    #include <iostream>
3    using namespace std;
4
5    int main()
6    {
7        Date arrayDate[4];              //array objects
```

8	Date	*day1 = new Date;	//*an object pointer*
9	Date	***day2 = new Date[4]**;	// *4 object pointers*
10	cout<<endl;		
11			
12	for(int i =0 ; i <4; i++)		
13	{		
14	cout<<"arrayDate["<<i<<"]'s date is ";		
15	**arrayDate[i].print();**		// *use member function*
16	cout<<"point to array of day["<<i<<"] date is ";		
17	**day2[i].print();**		
18	}		
19	cout<<endl;		
20			
21	delete day1;		//*free object day1*
22	delete []day2;		//*free 4 objects day2*
23	return 0;		
24	}		

Result:

1 ArrayDate[0] is constructed
2 ArrayDate[1] is constructed
3 ArrayDate[2] is constructed
4 ArrayDate[3] is constructed
5 ArrayDate[4] is constructed
6 ArrayDate[5] is constructed
7 ArrayDate[6] is constructed
8 ArrayDate[7] is constructed
9 ArrayDate[8] is constructed
10
11 arrayDate0[0]'s date is 1/1/2000
12 point to array of day[0] date is 1/1/2000
13 arrayDate0[1]'s date is 1/1/2000
14 point to array of day[1] date is 1/1/2000
15 arrayDate0[2]'s date is 1/1/2000
16 point to array of day[2] date is 1/1/2000
17 arrayDate0[3]'s date is 1/1/2000
18 point to array of day[3] date is 1/1/2000
19
20 ArrayDate[8] is destroyed
21 ArrayDate[7] is destroyed
22 ArrayDate[6] is destroyed
23 ArrayDate[5] is destroyed

24 ArrayDate[4] is destroyed
25 ArrayDate[3] is destroyed
26 ArrayDate[2] is destroyed
27 ArrayDate[1] is destroyed
28 ArrayDate[0] is destroyed

The beginning of this section stated that if you declare an array of class objects and class has constructor(s), then the class should have the default constructor, e.g. **Date()** in the header file (date.h) in Example 4-9.

4.7.2 Initialize an Array of Objects by Using Constructors with Parameters

Suppose that class **Date** has a constructor with parameters instead of a default constructor in the header file (**Date.h**) of Example 4-9, such as the following:

```
1   //Date.h
2
3   class Date{
4   public:
5       Date(int, int, int);
6       ~Date();
7       void print() const;
8   private:
9       int day, month, year;
10      static int count;
11  };
```

then the class object **arrayDate** is declared in the user program (e.g., example4_9.cpp) and initialized in the following way:

Date arrayDate[4] = { Date(6,4,1900),
                     Date(6,4,2000),
                     Date(15, 9, 2003),
                     Date(23, 7, 2010)};

In fact, the expression **Date(6, 4, 1900)** creates an anonymous object of the class **Date**; initializes its date members to 6, 4, 1900, respectively; and then uses member-wise copy to initialize the object **arrayDate[0]**.

4.8 Friends

As discussed in the previous sections on access specifiers, to access data members **day**, **month** and **year** from outside the class by the *main* function, the data members are declared as private inside the class **Date** in the previous examples, because the *main* function is not a member that will not be able to access the private data. To access private data members, data members are put into the public part of the class. However, this breaks the information hiding of the class. In fact, a programmer may have a situation where he or she would need to access private data from non-member functions. For handling such cases, the concept of **Friend** functions is a useful tool.

> A *friend* of a class is a function or class that is not a member of the class, but is granted the same access to the class as the members of the class.

Functions declared with the *friend* specifier in a class member list are called *friend functions* of that class. Classes declared with the *friend* specifier in the member list of another class are called *friend classes* of that class.

Difference between Ordinary Member Functions and Friend Functions

An ordinary **member function** of a class specifies three logically distinct things:
- the function can access the private part of the class,
- the function is in the scope of the class,
- the function must be invoked on the an object (has a *this* pointer).

A **friend** function of a class is considered as follows:
- the function is a non-member function of the class,
- the function can access to all the members (public or private) of the class like the member functions,
- the function declaration can be placed in either the private or the public part of a class declaration.

4.8.1 Friend Functions

The *friend* specifier appears only in the function prototype in the class definition, not in the definition of the friend function. The statement is as follows:

```
class class_name {
    //...
    friend returnValueType function_name (parameter list);
    //
};
```

The friend function must be defined outside the class because it is a non-member function of the class. Its definition is as follows:

```
returnValueType function_name (parameter list)
{ }
```

Example 4-10: A friend function.

```
//--------------------------------------------------------------------------
// File: Date.h
```
// This program defines a friend function within the **Date** class.

//--
1   class Date{
2   public:
3   Date(int, int, int);
4   //add n to member year
5   **friend void add_year (Date&, int);**   // *declaration of the friend function*
6   void print() const;
7   private:
8   int day, month, year;
9   };
//--
// File: **Date.cpp**
// The program presents the implementation detail of the **Date** class.
//--
1   #include "Date.h"
2   #include <iostream>
3   using namespace std;
4
5   Date::Date(int d, int m, int y) : day(d), month(m), year (y) {}
6   void Date::print() const
7   {
8   cout<<day<<month<<year<<endl;
9   }
10
11   //definition of the friend function
12   **void add_year (Date& d, int n)**   // *definition of friend function*
13   {
14   d.year += n;
15   }
//--
// File: **example4_10.cpp**
// This program demonstrates how to use a friend function.
//--
1 #include "Date.h"
2 #include <iostream>
3 using namespace std;
4 int main()
5 {
6   Date   today(19, 7, 2011);
7   today.print();

```
7      add_year(today, 5);    //invoke a friend function
8      today.print();
9      return 0;
10 }
```

In Example 4-10, data member **year** is declared in the *private* part of class **Date**. Friend function *add_year* has directly access to private member **year** of the class, although a friend function seems an ordinary function.

Sometimes a friend function can be a member of another class. For example,

Example 4-11: A friend function as a member of another class.

//---
// File: **example4_11.cpp**
// The program shows friend function *print* in class **A** to be a member of class **B**.
//---

```
1      #include <iostream>
2      using namespace std;
3
4      class A;
5
6      class B{
7      public:
8           void print(A& a);
9      };
10
11     class A{
12     public:
13          A() : x(1),y(2){}
14     private:
15          int x, y;
16          friend void B::print(A& a);      // friend function of class A
17     };
18
19     void B::print(A& a)
20     {
21          cout<<"x is "<<a.x<<endl;
22          cout<<"y is "<<a.y<<endl;
23     }
24
25     int main()
26     {
27          A aObj;
```

```
28      B bObj;
29      bObj.print(aObj);
30
31      return 0;
32  }
```
Result:
```
1   x is 1
2   y is 2
```

Here the class **A** must be defined before the member function *print* of the class **B** can be a friend of class **A**.

4.8.2   Friend Classes

You can declare an entire class as a friend. Suppose class **B** is a friend of class **A**. This means that every member function and static data member defined in class **B** has access to class **A**.

Example 4-12: A friend class.

//--
// File: **example4_12.cpp**
// The program shows friend class **B** declared in class **A**.
//--
```
1   #include <iostream>
2   using namespace std;
3
4   class A{
5   public:
6       A():x(1),y(2) {}
7   private:
8       int x, y;
9       friend class B;          // a friend class of class A
10  };
11
12  class B{
13  public:
14      void print(A& a);
15  };
16
17  void B::print(A& a)
18  {
19      cout<<"x is "<<a.x<<endl;
20      cout<<"y is "<<a.y<<endl;
21  }
```

```
22
23   int main()
24   {
25       A aObj;
26       B bObj;
27       bObj.print(aObj);
28       return 0;
29   }
```

Result:

```
1    x is 1
2    y is 2
```

As we see, the friend class **B** has a member function *print* that accesses the private data members **x** and **y** of class **A** and performs the same task as the friend function *print* in Example 4-11. Any other members declared in class **B** also have access to all members of class **A**.

> **Think These Over**
> 1. What are the differences among member functions, non-member functions and friend functions?
> 2. What is the purpose of using friend functions?

4.9　Case Study: Examples of Used-defined Types

4.9.1　A Better Date Class

The previous sections discussed pieces of design of the class **Date** in the context of introducing the basic language features for defining class. The design of a simple and efficient **Date** class is discussed again.

Generally, small, heavily-used abstractions are common in many applications. C++ and other programming languages directly support a few of these abstractions. However, most are not, and cannot be, supported directly because there are too many of them. Furthermore, the designer of a general-purpose programming language cannot foresee the detailed needs of every application. Consequently, mechanisms must be provided for the user to define small concrete types. Such types are called concrete types or concrete classes to distinguish them from abstract classes (see §7.3) and classes in class hierarchies (see §6).

To ensure a user-defined type, e.g. class **Date,** to be correct, readable and understandable, a few set of operations is considered:

● A constructor specifying how objects of the type are to be initialized and examining the validation of initialized data members.

● A set of functions allowing a user to examine a **Date**. These functions are marked const to indicate that they do not modify the state of the object for which they are called.

- A set of functions allowing the user to manipulate **Date** without actually having to know the details of the representation or fiddle with the intricacies of the semantics.
- A set of implicitly defined operations to allow **Dates** to be freely copied.

According to the above operations, a better **Date** class is built.

Example 4-13: Defintion of a better **Date** class.

//--
// File: **Date.h**
// This program describes the definition of a **Date** class.
//--

```
1    class Date
2    {
3      public:     //public interface
4
5        Date(int, int, int);            //constructor
6        Date(const Date&);              //copy constructor
7
8        //functions for changing the Date
9        void addDay(int);               //add n days
10       void setDate(int, int, int);    //modify a date
11
12       //functions for examining the Date.
13       int getDay() const;
14       int getMonth() const;
15       int getYear() const;
16
17       ~Date();                        //destructor
18
19     private:
20       int max(int, int);
21       int min(int, int);
22       bool IsRight(int, int, int);
23       bool IsLeapYear(int);
24
25       int year;
26       int day;
27       int month;
28       static int dayOfMonth[13];
29    };
```

The implementation for each member function is defined in the following:

//--
// File: **Date.cpp**
// The program presents the implementation detail of the **Date** class.
//--

```
1    #include "Date.h"
2    #include <iostream>
3    using namespace std;
4
5    int Date::dayOfMonth[] = {0,31,28,31,30,31,30,31,31,30,31,30,31};
6    //constructor
7    Date::Date(int d, int m, int y)
8    {
9        if(IsRight(d, m, y))
10       {
11           day = d;
12           month = m;
13           year = y;
14       }
15       else
16       {
17           year = max(1, y);
18
19           month = max(1, m);
20           month = min(month, 12);
21
22           dayOfMonth[2] = 28 + IsLeapYear(year);
23           day = max(1,d);
24           day = min(day, dayOfMonth[month]);
25       }
26   }
27   //copy constructor
28   Date::Date(const Date& d)
29   {
30       day = d.day;
31       month = d.month;
32       year = d.year;
33   }
34   //destructor
```

```
35    Date::~Date()
36    {
37    }
38    //modify a date
39    void Date::setDate(int d, int m, int y)
40    {
41        if(IsRight(d,m,y))
42        {
43            day = d;
44            month = m;
45            year = y;
46        }
47        else
48        {
49            year = max(1, y);
50
51            month = max(1, m);
52            month = min(month, 12);
53
54            dayOfMonth[2] = 28 + IsLeapYear(year);
55            day = max(1,d);
56            day = min(day, dayOfMonth[month]);
57        }
58    }
59    //add n days
60    void Date::addDay(int n)
61    {
62        while(1)
63        {
64            if(n + day >= dayOfMonth[month])
65            {
66                n = n + day - dayOfMonth[month];
67                month++;
68                if(month > 12)
69                {
70                    month = 1;
71                    year++;
72                    dayOfMonth[2] = 28 + IsLeapYear(year);
73                }
```

```
74              day = 0;
75          }
76          else break;
77      }
78      day = n;
79  }
80
81  int Date::getDay() const
82  {
83      return day;
84  }
85  int Date::getMonth() const
86  {
87      return month;
88  }
89  int Date::getYear() const
90  {
91      return year;
92  }
93  //examine a leap year
94  bool Date::IsLeapYear(int y)
95  {
96      return !(y%4) && (y%100) || !(y%400);
97  }
98  //examine a valid date
99  bool Date::IsRight(int d, int m, int y)
100 {
101     int maxDay;
102
103     if((y>9999) || (y<1) || (d<1) || (m<1) || (m>12))
104         return false;
105
106     maxDay = 31;
107     switch(m)
108     case 4: case 6: case 9: case 11: maxDay--;
109     if(m == 2) maxDay = IsLeapYear(y) ? 29 : 28;
110     return (d>maxDay)? false : true;
111 }
112
```

```
113     int Date::max(int a, int b)
114     {
115         if(a>b) return a; return b;
116
117     }
118
119     int Date::min(int a, int b)
120     {
121         if(a<b) return a;return b;
122     }
```

In the above program, the constructor checks that the data supplied denotes a valid Date by a call of function *IsRight()*. If not, the date is set to Date(1,1,1). If the data supplied is acceptable, the obvious initialization is done. Initialization is a relatively complicated operation because it involves data validation. This is fairly typical. On the other hand, once a Date has been created, it can be used and copied without further checking.

As is common for such simple concrete types, the definitions of member functions vary between the trivial and the not-too-complicated.

Additionally, several member functions in the ***private*** part of class **Date** are defined. For example,

        int max(int, int);
        int min(int, int);
        bool IsRight(int, int, int);
        bool IsLeapYear(int);

These functions are associated with the class **Date**, but they need not to be defined in the class itself because they do not need direct access to the representation, that is, they are not accessed directly by its objects.

Defining such functions in the public part of the class would complicate the class interface and increase the number of functions that would potentially need to be examined when a change to the representation was considered.

Finally, class **Date** is tested as follows:

```
//-----------------------------------------------------------------------
// File: example4_13.cpp
// The program shows how to use the Date class.
//-----------------------------------------------------------------------
1    #include "Date.h"
2    #include <iostream>
3    using namespace std;
4
5    int  main()
```

```
6    {
7         Date d(26, 3, 1990);
8         cout<<d.getDay()<<"/"<<d.getMonth()<<"/"<<d.getYear()<<endl;
9
10        d.setDate(45, 45, 1990);
11        d.addDay(3);
12        cout<<d.getDay()<<"/"<<d.getMonth()<<"/"<<d.getYear()<<endl;
13
14        Date d1 = d;
15        d1.addDay(20);
16        cout << d1.getDay() << "/"<<d1.getMonth() << "/" << d1.getYear() <<endl;
17
18        return 0;
19   }
```

Result:

1   26/3/1990

2   3/1/1991

3   23/1/1991

Now we finish the definition of the class **Date** and its implementation. The class **Date** is a simple small user-defined type. A good set of such types, like **Date**, can provide a foundation for applications. Lack of suitable "small efficient types" in an application can lead to gross run-time and space inefficiencies when overly general and expensive classes are used. Alternatively, lack of concrete types can lead to obscure programs and time wasted when each programmer writes code to directly manipulate "simple and frequently used" data structures.

4.9.2   A GradeBook Class with Objects of the Student Class

In Section 3.8, we define a **GradeBook** class. As we known, a grade book is used to maintain a number of student grades. However, we do not define an attribute of the student grade in this class. Moreover, only a grade attribute is not enough to represent all meanings of a student. Suppose that we consider each student has the attributes of student ID, name and grade. Therefore, a data member we add into the GradeBook class is an aggregation of several built-in types, that is, an object of the **student** class type is considered as a member of the **GradeBook** class. Thus, "a grade book has a student" is an aggregation relationship between the **GradeBook** class and the **Student** class. The UML diagram of their relationship is shown in Figure 4-5. The empty diamond indicates aggregation relationship in Figure 4-5 (Notice that composition is a special case of the aggregation relationship). The underline indicates class static members.

Chapter 4   Classes and Objects (II)

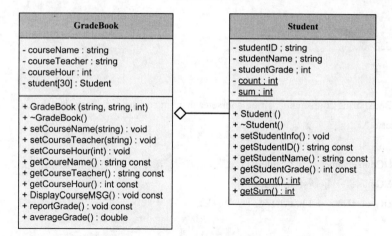

Figure 4-5   The UML diagram of the aggregation relationship between
the **GradeBook** class and the **Student** class.

Example 4-14: Demonstrate how to define a GradeBook class and use it.

//--
// File: **GradeBook.h**
// The program presents the declaration of two classes Student and GradeBook.
//--

```
1   const int MaxNumberOfStudent = 30;   // the maximum number of students
2   //Student class
3   class Student{
4   public:
5       Student();
6       ~Student();
7
8       void setStudentInfo();
9
10      string getStudentID() const;
11      string getStudentName() const;
12      int getStudentGrade() const;
13
14      static int getSum();
15      static int getCount();
16  private:
17      string studentID;           // student ID
18      string studentName;         // student name
19      int studentGrade;           // student grade
20      static int count;           // student number
21      static int sum;             // a sum of student grades
22  };
```

```
23  // GradeBook class
24  class GradeBook
25  {
26  public:
27      GradeBook(string coursename, string courseteacher, int coursehour, int studentNum);
28      ~GradeBook();
29
30      void setCourseName( string name );
31      void setCourseTeacher( string teacher );
32      void setCourseHour( int hour);
33
34      string getCourseName() const;
35      string getCourseTeacher() const;
36      int getCourseHour() const;
37      // display course messages to the GradeBook user
38      void displayCourseMSG() const;
39      // report all student grades
40      void reportGrade() const;
41      // calculate the average of students
42      double averageGrade();
43  private:
44      string courseName;        // course name
45      string courseTeacher;     // course teacher
46      int courseHour;           // course hour
47      int studentNumber;
48      Student *student;         // a number of students
49  };
```

//--
// File: **GradeBook.cpp**
// The program presents the implementation of classes Student and GradeBook.
//--

```
1   #include <iostream>
2   #include <string>
3   #include <iomanip>
4   #include <conio.h>
5   using namespace std;
6
7   #include "GradeBook.h"    // include definition of class GradeBook
8   // initialize static data members
9      int Student::count = 0;
10  int Student::sum = 0;
```

```
11  //implementation detail of the Student class
12  Student::Student()
13  {
14      studentID = " ";     studentName = " ";     studentGrade = 0;
15  }
16  Student::~Student()
17  {
18      if(count > 0)
19          cout << "Destroying student object" << count << endl;
20      --count;
21  }
22  void Student::setStudentInfo()
23  {
24      cout << " Enter student's ID, name and grade\n";
25      cin >> studentID >> studentName >> studentGrade;
26      count ++; sum += studentGrade;
27  }
28  string Student::getStudentID() const
29  { return studentID; }
30  string Student::getStudentName() const
31  {     return studentName;     }
32   int Student::getStudentGrade() const
33  { return studentGrade; }
34  int Student::getSum()
35  {   return sum;   }
36  int Student::getCount()
37  {   return count;   }
38
39  // implementation details of the GradeBook class
40  GradeBook::GradeBook(string coursename, string courseteacher, int coursehour, int studentNum)
41  {
42      setCourseName( coursename );              // call set function to initialize courseName
43      setCourseTeacher(courseteacher);
44      setCourseHour( coursehour);
45
46      studentNumber = studentNum;
47      student = new Student[studentNumber];     //declare the student objects
48
49      int i = 0;
46      while(1)
47      {
```

```cpp
48         student[i].setStudentInfo();
49         ++i;
50     }
51 }
52 GradeBook::~GradeBook()
53 {  cout << "Destroying GradeBook object\n";
54    delete []student;   }                    //destroy all student objects
55
56 void GradeBook::setCourseName( string name )
57 {  courseName = name; }                     // store the course name in the object
58 void GradeBook::setCourseTeacher( string teacher )
59 {  courseTeacher = teacher; }               // store the course name in the object
60 void GradeBook::setCourseHour( int hour)
61 {  courseHour = hour;   }
62
63 string GradeBook::getCourseName() const
64 {  return courseName; }                     // return object's courseName
65 string GradeBook::getCourseTeacher() const
66 {  return courseTeacher; }                  // return object's courseTeacher
67 int GradeBook::getCourseHour() const
68 {  return courseHour;   }                   // return object's courseHour
69
70 void GradeBook::displayCourseMSG() const
71 {
72     // call getCourseName to get the courseName
73     cout << "Welcome to the grade book for " << getCourseName() << "!" << endl;
74     cout << "Teacher: " << getCourseTeacher() << "; "
75          << "Hour: " << getCourseHour() <<endl;
76 }
77 void GradeBook::reportGrade() const
78 {
79     cout << "Student Grade Report\n";
80     cout << setw(10) << "ID" << setw(10) << "Name" << setw(4) << " Grade\n";
81     for(int i = 0; i < Student::getCount(); i++)
82         cout << setw(10) << student[i].getStudentID()
83              << setw(10) << student[i].getStudentName()
84              << setw(4) << student[i].getStudentGrade() <<endl;
85 }
86 double GradeBook::averageGrade()
87 {
88     return (double)Student::getSum() / Student::getCount();
```

89   }
//---
// File: **example4-14.cpp**
// The program tests the implementation of classes Student and GradeBook.
//---

```
1   #include <iostream>
2   using namespace std;
3   #include "GradeBook.h"
4   int main()
5   {
6       int number;
7       do
8       {
9           cout << "Enter the number of student:";
10          cin >> number;
11      } while (number > MaxNumberOfStudent);
12      // create a GradeBook object
13      GradeBook gradeBook( "Object-Oriented Programming in C++", "Christina", 48, number);
14      cout << "-----------------------------------------------------------------------\n";
15      gradeBook.displayCourseMSG();
16      cout << "-----------------------------------------------------------------------\n";
17      gradeBook.reportGrade();
18      cout << "-----------------------------------------------------------------------\n";
19      cout << "The average grade of students: " << gradeBook.averageGrade() << endl;
20      cout << "-----------------------------------------------------------------------\n";
21      return 0;
22  }
```

Input:

Enter the number of student:3
Enter student's ID, name and grade
1002301 wang 75
Enter student's ID, name and grade
1002302 liu 90
Enter student's ID, name and grade
1002301 zhao 65

Result:

```
1   -------------------------------------------------------------------------
2   Welcome to the grade book for Object-Oriented Programming in C++!
3   Teacher: Christina; Hour: 48
4   -------------------------------------------------------------------------
5   Student Grade Report
```

```
 6          ID          Name      Grade
 7       1002301        wang       75
 8       1002302        liu        90
 9       1002301        zhao       65
10  ------------------------------------------------------------
11  The average grade of students: 76.6667
12  ------------------------------------------------------------
13  Destroying GradeBook object
14  Destroying student object3
15  Destroying student object2
16  Destroying student object1
```

Word Tips

alternatively	*adv.* 或者		literal	*adj.* 符号的
application	*n.* 应用程序		obscure	*adj.* 费解的
assignment	*n.* 赋值		potential	*adj.* 有可能的
composition	*n.* 组合		recommend	*vt./vi.* 建议，推荐
duplicate	*vt./vi.* 副本，复制		shallow	*adj.* 浅的
dynamically	*adv.* 动态地		stack	*n.* 栈
foresee	*vt.* 预知		storage	*n.* 存储
impractical	*adj.* 不现实的		string	*n.* 字符串
increment	*n.* 增量		suffix	*n.* 后缀
instantiate	*vt.* 举例		violate	*vt.* 违反，违背
intricacy	*n.* 错综复杂			

Exercises

1. Mark the following statements as true or false and give reasons.

(1) A set function should perform validity examining.

(2) Assigning an object to an object of the same type results in default member-wise copy.

(3) Member functions declared const cannot modify the object.

(4) An object cannot be declared as const.

(5) A class cannot have objects of other classes as members.

(6) If a member initializer is not provided for a member object, the member object's default constructor is called implicitly.

(7) Static members are shared by all instances of a class.

(8) A class's static member only exists when an object of the class exists.

(9) Information hiding is a process by which a class's data implementation is hidden from the user of the class.

(10) The primary activity in C++ is creating objects from abstract date types and expressing

Chapter 4 Classes and Objects (II)

interactions between those objects.

(11) A friend function of a class can access all private data of the class.

(12) A friend function of a class is a member of the class.

2. Find the error(s) in each of the following and explain how to correct it.

(1)
```
class AA{
    private:
        int a;
        int b;
    public:
        AA( int x)
        {
            a = 0;
            b = x;
        )
        void print() const
        {   b++;
            cout<<b; }
    };
    void main()
    {    AA   aa;
        aa.print();
        cout<<AA::a;
    }
```

(2)
```
class A{
    int x;
    Date d;
    static int y;
    public:
    void A(int a)
    { x = a;
      d(a, a, a)
      y = 0;
    }
};
void main()
{    A aa;
    cout<<aa.y<<endl;
    A bb(10);
    cout<<bb.y<<endl;
}
```

135

3. Write out the output of the following codes:

(1)
```cpp
class A
{
        int a;
public:
        A()
            {
            a=0;
            cout<<"constructing default A"<<endl;
            }
            A(int x)
            {
            a = x;
            cout<<"constrcting A"<<endl;
            }
    ~A(){cout<<"destructing A"<<endl;}
};
class B
{
    A a;
public:
    B()
    {
        cout<<"constructing default B"<<endl;
    }
    B(int x):a(x)
    {
        cout<<"constructing B"<<endl;
    }
    ~B(){cout<<"destructing B"<<endl;}
};
int main()
{
    B b1;
    B b2(10);
    return 0;
}
```

(2)
```cpp
class Demo{
```

```cpp
        int n;
public:
    Demo(){}
    Demo(int m) { n=m; }
    friend Demo square(Demo s){
        Demo s1;
        s1.n = s.n * s.n;
        return s1;
    }
    void disp(){    cout<<n<<endl;    }
};
void main{
    Demo a(10);
    a=square(a);
    a.disp();
}
```

(3)
```cpp
class Demo{
    int n;
    static int sum;
public:
    Demo(int x){    n=x;    }
    void add(){    sum+=n;    }
    void disp(){    cout<<"n="<<n<<", sum="<<sum<<endl;    }
}
int Demo::sum = 0;
void main(){
    Demo a(2), b(3), c(5);
    a.add();        a.disp();
    b.add();        b.disp();
    c.add();        c.disp();
}
```

(4)
```cpp
class Count
{
public:
    int value;
    Count(const Count& c)
    {    value = c.value; cout <<"Copy an object\n"; }
```

```
            Count()
            {     value = 0; cout <<"Creating an object \n";     }
    };
    void increment (Count c, int time)
    {     cout <<"Calling the increment function\n";
        c.value++;
            time++;     }
    int main()
    {
            Count myCount;
            int times = 0;

            for (int i = 0; i < 3; i++)
                increment(myCount, times);

            cout << "myCount.count is " << myCount.value <<endl;
            cout << " times is " << times <<endl;
            return 0;
    }
```
if the statement *void increment(Count c, int times)* is changed to

    void increment(Count& c, int& times),

what will be the output?

4. Write a program that the class **Cube** is defined with the properties **Length**, **Width**, **Height**. The class includes the following definition:

   (1) Two constructors, a copy constructor and destructor;
   (2) A member function of calculating the volume of Cube;
   (3) A member function of displaying the properties;
   (4) A member function of moving a Cube;
   (5) A member function of modify the properties of Cube.

   Test the class **Cube** by using a main function.

5. Define a class called **employee** that contains a **name** (an object of class string) and an employee **number** (type long). Include a member function called *getdata* to get data from the user for insertion into the object, and another function called *putdata* to display the data. Assume the name has no embedded blanks. Write a main program to exercise this class. It should create an array of type employee, and then invite the user to input data for up to 100 employees. Finally, it should print out the data for all the employees.

6. Define a **Dictionary** class that has **Word** objects defined in the previous execise and the number of words. Include member function *FindWord* to find a word, member function *AddWord* to add a word, member function *GetWord* to obtain a word and member function

Print to output all word.
7. Write the definition of three classes **Bank1**, **Bank2** and **Bank3**. These three classes all have private data member **balance** and member function *display* to output the balance. Design a friend function *total* in the three classes to calculate the total balance from the three banks. Test the program in a *main* function.

Chapter 5
Operator Overloading

When I use a word it means just what I choose it to mean —
neither more nor less.
—Humpty Dumpty

Objectives
- To learn about operator overloading
- To explore the member and non-member operator functions of a class
- To understand how to overload operators to work with user-defined types
- To understand how to convert objects from one class to another class

5.1 Why Operator Overloading Is Need

The previous chapter defined and implemented the class **Date**. It also showed how the class **Date** can be used to represent a date in the user's program. Let's review some of definition of the class **Date**.

Two objects are declared as follows:

Date myDay(12, 6, 2000);

Date yourDay(23, 9, 2005);

Now we do the following operations:

myDay.print();

myDay.add_year(1);

if(myDay.equalTo(yourDay))

……

The first statement prints the value of the object **myDay**. Whether can we use the insertion operator << to output the value of **myDay**? The second statement increments the value of **myDay** by one year, whether can we use the increment operator ++ to increment the value of **myDay**? Likewise, whether can we use relational operator for comparison? If yes, we can enhance flexibility for manipulating object **myDay**. Thus, we can use the following statements instead of the previous statements:

cout<<myDay;

myDay++;

if (myDay == yourDay)

……

Chapter 5 Operator Overloading

As we know, C++ supports a set of operators for built-in types. However, these operators cannot directly be applied to user-defined types. Although we want to use the operators <<, ++ and == to represent the notions of the object **myDay**, this will cause compiling errors since we do not define the behavior of the class **Date** with <<, ++ and ==. Therefore, we must extend the definition of these operators in our class **Date**. This is called *operator overloading*, in C++ terminology. Operator overloading is one of the most exciting features of object-oriented programming.

5.2 Operator Functions

5.2.1 Overloaded Operators

C++ allows the user to overload most of operators so that the operations can work effectively in a specific application. Here is a list of all the operators that can be overloaded:

```
+    -    *    /    =    <    >    +=   -=   *=   /=   <<   >>   <<=  >>=
==   !=   <=   >=   ++   --   %    &    ^    !    |    ~    &=   ^=   |=
&&   ||   %=   []   ()   ,    ->*  ->   new  delete  new[]  delete[]
```

Actually, C++ does not allow to overload all operators. The following operators cannot be defined by a user:

:: (scope resolution)
. (member selection)
.* (member selection through pointer to member)
?: (ternary operator)

5.2.2 Operator Functions

In order to overload operators, you must write functions. The name of the function that overloads an operator is reserved word operator followed by the operator to be overloaded.

| The *operator function* is a function that overloads an operator. |

The format of the operator function is:

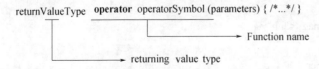

First, we give an example of the class complex number. A number of the form $a + ib$, where $i^2 = -1$, and a and b are real numbers, is called a *complex number*.

Example 5-1: A member operator function and a non-member operator function.

```
//--------------------------------------------------------------------------------
// File: complex.h
```

141

// This program describes the definition of a **complex** class in which there is a member operator (+) function.
//--

```
1    class complex{
2    public:
3        complex(double, double);
4        //operator function as a member
5        complex operator + (complex&);        //overloading operator +
6        void print() const;
7
8        double real, imag;
9    };
```
//--
// File: **complex.cpp**
// The program presents the implementation detail of the **compex** class and the definition of a non-
// member operator (-) function.
//--

```
1    #include "complex.h"
2    #include <iostream>
3    using namespace std;
4
5    complex::complex(double re, double im) : real( re ), imag( im ){}
6
7    complex complex::operator + (complex& c)        //overloading operator+
8    {
9        complex temp(0, 0);
10       temp.real = real + c.real;
11       temp.imag = imag + c.imag;
12       return temp;
13   }
14
15   void complex::print() const
16   {
17       cout<<real<<"  +  "<<imag<<"i\n";
18   }
19   //operator function as a non-member
20   complex operator – (complex& c1, complex& c2)
21   {
22       complex temp(0, 0);
23       temp.real = c1.real   – c2.real;
24       temp.imag = c1.imag – c2.imag;
25       return temp;
```

26   }
//---
// File: **example5_1.cpp**
// This program shows how to overload operators + and −.
//---
```
1      #include "complex.h"
2      #include <iostream>
3      using namespace std;
4
5      int main()
6      {
7          complex x(2.3, 4.5), y(3.0, 1.2);
8
9          complex z = x + y;      // shorthand : equivalent to z=x.operator + (y)
10         z.print();
11
12         z = x − y;              //z = operator − (x, y)
13         z.print();
14
15         return 0;
16     }
```
Result:
```
1      5.3 + 5.7i
2      −0.7 + 3.3i
```

Example 5-1 presents two operator functions of addition operator (+) and minus operator (−). They are defined as follows:

    complex complex::operator + (complex& c)

    complex operator − (complex& c1, complex& c2)

The first function overloads an addition operator (+) as a member function. The second function overloads a subtraction operator (-) as a non-member function. Notice that we have to access data members **real** and **imag** in the non-member operator function so we put these data members into the public part of the class.

However, putting data members **real** and **imag** into the public part of class **complex** breaks the information hiding of the class. We can define a subtraction operator (-) as a friend function.

Example 5-2: A friend operator function.

//---
// File: **complex.h**
// This program describes the definition of a **complex** class in which there is a friend operator (−) function.
//---

```
1    class complex{
2    public:
3           complex(double, double);
4           //operator function
5           friend complex operator - (complex&, complex&);
6           void print() const;
7    private:
8           double real,   imag;
9    };
```
//--
// File: **complex.cpp**
// The program presents the implementation detail of the **complex** class and the definition of a friend
// operator (-) function.
//--
```
1    #include "complex.h"
2    #include <iostream>
3    using namespace std;
4
5    complex::complex(double re, double im) : real( re ), imag( im ){}
6
7    void complex::print() const
8    {
9         cout<<real<<" +"<<imag<<"i\n";
10   }
11   //definition of the friend function
12   complex operator - (complex& c1, complex& c2)
13   {
14
15       complex temp(0, 0);
16       temp.real = c1.real   - c2.real;
17       temp.imag = c1.imag - c2.imag;
18       return temp;
19 }
```

The user program is the same as that of Example 5-1. In Example 5-2, data members **real** and **imag** are declared in the *private* part of class **complex**. Friend function *operator -* has directly access to these private members of the class. Thus, a friend function is an ordinary function.

The above gives an example about that the + and - operators are overloaded as a member function and non-member function. So, when can the operators be overloaded as either member functions or non-member functions?

As we know, the + operator is overloaded as a member function. The + operator has direct access to the data members of the objects, and you need to pass only one object as a parameter. On the other hand, when the - operator is overloaded as a non-member function, then you must pass both objects as parameters. Therefore, overloading operators as a non-member could require additional memory and computational time to make a local copy of the data. Thus, for efficiency purpose, wherever possible, you should overload operators as member functions.

When overloading an operator, keep the following in mind:
- You cannot change the precedence of an operator.
- The associativity cannot be changed. For example, the associativity of arithmetic operator addition is from left to right, and it cannot be changed.
- Default parameters cannot be used with an overloaded operator.
- You cannot change the number of parameters an operator takes.
- You cannot create new operators. Only existing operators can be overloaded.
- The following operators cannot be overloaded:
    .   .*   ::   ?:   sizeof
- The meaning of how an operator works with built-in types, such as int, remains the same.
- Operators can be overloaded either for objects of the user-defined types, or for a combination of objects of the user-defined types and objects of the built-in type.

Think These Over
1. Why do we need to overload the operators?
2. When can the operators be overloaded as either member functions or non-member functions?

5.3 Binary and Unary Operators

5.3.1 Overloading Binary Operators

In general, arithmetic, such as +, -, *, or relation, such as ==, <= represent a binary operator. These **binary operators** can be defined by either a non-static *member function* taking one argument or a *nonmember (or friend) function* taking two arguments.

Overloading Binary Operators as Member Functions

If the operator function is a member of the class, it has one argument. The declaration of its prototype is

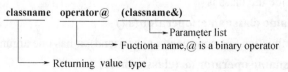

and its definition outside the class is

    classname   classname::operator@(classname&)

For example,

```
1  class complex{
2  public:
3      complex(double, double);
4
5      //overloading a binary operator +
6      complex operator + (complex&);    //member function taking one argument
7  private:
8      double real,   imag;
9  };
```

Overloading Binary Operators as Non-Member Functions

If the operator function is a non-member (or friend) of the class, it has two arguments. Its definition is

    classname   operator @   (classname&, classname&)

For example,

```
1  //non-member function
2  complex operator – (complex& c1, complex& c2)   //taking two arguments
3  {
4      complex temp(0, 0);
5      temp.real = c1.real   – c2.real;
6      temp.imag = c1.imag – c2.imag;
7      return temp;
8  }
```

5.3.2   Overloading Unary Operators

The process of overloading unary operators is similar to the process of overloading binary operators. The only difference is that in the case of unary operators, the operator has only one argument. Therefore, to overload a unary operator @ for a class:

• If the operator function is a member of the class, it has no argument. The declaration of its prototype is

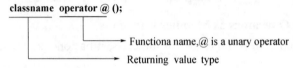

and its definition outside the class is

    classname classname::operator @()

• If the operator function is a non-member (or friend), it has one argument. Its definition is

    classname operator @ (classname&)

Chapter 5 Operator Overloading

Now we describe how to overload two unary operators, that is, increment (++) and decrement (--)operators.

The increment operator has two forms: **prefix** increment (++a) and **postfix** increment (a++), where **a** is a type variable, such as a type *int*, or a user-defined type. In the case of prefix increment, the value of **a** is incremented by 1 before the value is used in the expression. In the case of postfix increment, the value of **a** is used in the expression before it is incremented by 1.

Overloading Prefix Increment Operator(++)

Suppose that **c** is an object of class **complex**, the statement:
```
++c;
```
increments the values of **real** and **imag** of **complex** by 1.

Thus, overloading an increment operator (++) as a member of the class **complex** is defined as follows:

```
1    class complex{
2    public:
3            complex(double, double);
4
5    // overloading a unary ++ operator as member function
6            complex& operator ++ ();    //member function taking no argument
7    private:
8            double real, imag;
9    };
10
11   complex& complex::operator++()    //definition of member function
12   {
13           real++;
14           imag++;
15           return *this;
16   }
```

The following example is that overloading operator the ++ as a friend of the class **complex**.

```
1    class complex{
2    public:
3            complex(double, double);
4
5            // overloading a unary ++ operator as non-member function
6            friend complex operator ++ (complex&);    // taking one argument
7    private:
8            double real,   imag;
9    };
10
11   complex operator++(complex& c)    //definition of friend function
```

```
12    {
13        (c.real)++;
14        (c.imag)++;
15        return c;
16    }
```

Overloading Postfix Increment Operator

The above examples present the prefix increment operator (++) to be overloaded. To distinguish between pre- and post-fix increment operator overloading, we use a dummy parameter (of type int) in the function declaration of the operator function. Thus, the function prototype for the postfix increment operator of the class **complex** is:

- **complex& operator++(int)**   for the member function, or
- **complex operator++(complex&, int)**   for the non-member function (or friend)

Suppose that **c** is an object of class **complex**, the statement:

   c++;

is compiled by compiler in the statement: c.operator++(0).

Example 5-3: The **complex** class with overloading prefix and postfix operator (++).

//--
// File: **complex.h**
// This program describes the definition of a **complex** class with the friend function of prefix
// increment operator and a member function of postfix increment operator.
//--

```
1     class complex{
2     public:
3         complex( double =0, double =0);
4
5         //non-member function
6         friend complex operator ++ (complex&);      //prefix increment operator
7
8         //member function
9         complex operator ++ (int);                  //postfix increment operator
10
11        void print() const;
12    private:
13        double real, imag;
14    };
```
//--
// File: **complex.cpp**
// The program presents the implementation detail of the **compex** class.
//--

```
1    #include "complex.h"
2    #include<iostream>
3    using namespace std;
4
5    complex::complex(double re, double im) : real(re), imag(im) {}
6
7    //member function
8    complex complex::operator ++(int)
9    {
10           complex temp = *this;
11           real++;
12           imag++;
13           return temp;
14   }
15   //non-member function
16   complex operator++ (complex& c)
17   {
18           c.real++;
19           c.imag++;
20           return c;
21   }
22
23   void complex::print() const
24   {
25           cout<<real<<"+ "<<imag<<"i\n";
26   }
```
//---
// File: **example5_3.cpp**
// This program shows how to overload prefix and postfix operators ++.
//---
```
1    #include "complex.h"
2    #include<iostream>
3    using namespace std;
4
5    void main()
6    {
7         complex c1(20.3,11.3);
8
9         cout << "Original value of c1: ";
10        c1.print();
11
```

```
12      ++c1;                           // prefix increment
13      cout << "Value after ++c1: ";
14      c1.print();
15
16      c1++;                           // postfix increment
17      cout << "Value after c1++: ";
18      c1.print();
19
20      complex c2 = ++c1;
21      cout << "\nValue of c1 after c2 = ++c1: ";
22      c1.print();
23      cout << "Value of c2 after c2 = ++c1: ";
24      c2.print();
25
26      c2 = c1++;
27      cout << "\nValue of c1 after c2 = c1++:";
28      c1.print();
29      cout << "Value of c2 after c2 = c1++: ";
30      c2.print();
31  }
```

Result:

```
1  Original value of c1:   20.3 + 11.3i
2  Value after ++c1:   21.3 + 12.3i
3  Value after c1++:   22.3 + 13.3i
4
5  Value of c1 after c2 = ++c1:   23.3 + 14.3i
6  Value of c2 after c2 = ++c1:   23.3 + 14.3i
7
8  Value of c1 after c2 = c1++:   24.3 + 15.3i
9  Value of c2 after c2 = c1++:   23.3 + 14.3i
```

Think It Over

Why do the operator functions take different paraments between member and non-member functions?

5.4 Overloading Combinatorial Operators

The meanings of some built-in operators are defined to be equivalent to some combination of other operators on the same arguments. For example, if **a** and **n** are integer, **a += n** means **a = a + n**. Such relations do not hold for user-defined operators unless the user happens to define them that way. For example, a compiler will not generate a definition of **complex::operator+=()** from the definitions of **complex::operator + ()** and **complex::operator = ()**. Therefore, you must predefine an addition assignment operator (+=).

Example 5-4: Overloading combinatorial operator (+=).

```
//----------------------------------------------------------------------
// File: complex.h
// This program describes the definition of a complex class with overlaoding operator +=.
//----------------------------------------------------------------------
1    class complex{
2    public:
3        complex(double =0, double =0);        //constructor
4        complex(const complex&);              //copy constructor
5
6        //member function
7        complex operator + (complex&);        //overloading addition +
8        complex& operator = (complex&);       //overloading assignment =
9
10       complex& operator += (complex&);      //predefining operator +=
11       void print() const;
12   private:
13       double real, imag;
14   };
//----------------------------------------------------------------------
// File: complex.cpp
// The program presents the implementation detail of the compex class.
//----------------------------------------------------------------------
1    #include "complex.h"
2    #include<iostream>
3    using namespace std;
4
5    complex::complex(double re, double im) : real(re), imag(im)
6    {
7        cout<<"Constructing an object\n";
8    }
9
10   complex::complex(const complex& c)
11   {
12       real = c.real;
13       imag = c.imag;
14       cout<<"Copying an object\n";
```

```
15    }
16
17    complex complex::operator +(complex& c)
18    {
19          complex temp;
20          temp.real = real + c.real;
21          temp.imag = imag + c.imag;
22          return temp;
23    }
24
25    complex& complex::operator =(complex& c)
26    {
27            real = c.real;
28            imag = c.imag;
29            cout<<"Overloading assignment operator =\n";
30            return *this;
31    }
32
33    complex& complex::operator += (complex& c)
34    {
35            *this = *this + c;
36            return *this;
37    }
38
39    void complex::print() const
40    {
41          cout<<real<<" + "<<imag<<"i\n";
42    }
```
//---
// File: **example5_4.cpp**
// This program shows how to overload operator +=.
//---
```
1    #include "complex.h"
2    #include<iostream>
3    using namespace std;
4
5    void main()
6    {
7          complex c1(20.3,11.3);
8          cout << "Original value of c1: ";
9          c1.print();
```

10
11   complex c2 = c1;
12   cout << "Original value of c2: ";
13   c2.print();
14
15   complex c3;
16   cout<<"value of c3 after c3 = c1 is \n";
17   c3 = c1;
18   c3.print();
19   cout<<"value of c3 after c3 += c2 is ";
20   c3 += c2;
21   c3.print();
22   }

Result:

1 Constructing an object
2 Original value of c1: 20.3 + 11.3i
3 Copying an object
4 Original value of c2: 20.3 + 11.3i
5 Constructing an object
6 value of c3 after c3 = c1 is
7 Overloading assignment operator =
8 20.3 + 11.3i
9 value of c3 after c3 += c2 is 40.6 + 22.6i

You may notice that the two statements in Lines 11 and 17 (in example5_4.cpp) output the different results. These two statements look very similar. However, they have the different meanings. The statement in Line 11 initializes an object **c2** by constructed object **c1** and invokes the copy constructor of class **complex**, copying into it the members from **c1**. The output is shown in Line 3 of Result. The statement in Line 17 invokes the member function of assignment operator (=) of class complex, copying into it the members from **c1**. The output is shown in Line 7 of the result.

Copy Constructor and Assignment Operator

Sometimes, the default copy constructor and default assignment operator are used in the program. If we do not define a copy constructor and an assignment operator in the file **complex.h** of Example 5-4, we use the following statements:

   complex c2 = ++c1;
   c2 = c1++;

The first statement invokes a default copy constructor to initialize the object **c2** by **++c1**. The second statement invokes a default assignments operator to copy **c2** the new data from **c1++**. The situation with the default implementations is that a simple copy of the members (i.e. *Shallow Copy*) may not be appropriate to clone an object.

Object-Oriented Programming in C++

 The difference between *copy constructor* and *assignment operator* is that the copy constructor of the target is invoked when the source object is passed in at the time the target is constructed, such as in Line 11. The assignment operator is invoked when the target object already exists, such as in Line 17.

For instance, what if one of the members was a pointer that is allocated by the class? The shallow copying the pointer is not enough because now you'll have two objects that have the same pointer value, and both objects will try to free the memory associated with that pointer when they destruct. In this case, we must define a copy constructor and an assignment operator for copying objects (i.e. *Deep Copy*).

5.5 Mixed Arithmetic of User-Defined Types

An operator function must either be a member or take at least one argument of a user-defined type (functions redefining the *new* and *delete* operators need not). This rule ensures that a user cannot change the meaning of an expression unless the expression contains an object of a user-defined type. In particular, it is not possible to define an operator function that operates exclusively on pointers. This ensures that C++ is extensible but not mutable (with the exception of operators =, &, and, for class objects).

An operator function intended to accept a basic type as its first operand cannot be a member function. For example,

Complex c(2.4, 3.5);

Complex x = c + 2;

Complex y = 2 + c;

The second line states that a complex object **c** is added to the integer **2**: **c + 2** can, with a suitably declared member function, be interpreted as *a.operator* + (2), that is,

c+2 $\Longleftrightarrow$ c.operator + (2)

However, the **2 + c** statement in the third line cannot be done because there is no class **int** for which to define operator + to mean 2.*operator* + (c), that is,

2 + c $\Longleftrightarrow\!\!\!/$ 2.operator + (c)

Even if there were, two different member functions would be needed to cope with **2 + c** and **c + 2**. Because the compiler does not know the meaning of a user-defined +, it cannot assume that it is commutative and so interpret **2 + c** as **c + 2**.(see Page142).

 Think It Over

How do you implement the operations of **2 + c** and **c + 2** for a class object **c**?

5.6 Type Conversion of User-Defined Types

Most programs process information of a variety of types. Sometimes all the operations "stay within a type". For instance, adding an integer to an integer produces an integer. But it is

154

often necessary to convert data of one type to data of another type. This can happen in assignments, in calculations, in passing values to functions, and in returning values from functions. The compiler knows how to perform certain conversion among built-in types. Programmers can force conversions among built-in types by casting.

But what about user-defined types? The compiler cannot automatically know how to convert among user-defined types and built-in types. Therefore, such conversion can be perform by a constructor, called *conversion constructors*, namely, single argument constructors that convert objects of other types (including built-in types) into objects of a particular class.

Using a constructor to specify type conversion is convenient but has implications that can be undesirable. A constructor cannot specify

- an implicit conversion from a user-defined type to a basic type (because the basic types are not classes), or
- a conversion from a new class to a previously defined class (without modifying the declaration for the old class).

A *conversion operator* (also called a *cast operator*) can be used to convert an object of one class into an object of another class or into an object of a built-in type. Such a conversion operator must be a non-static member function, but not a friend function. The function of conversion operator is as follows:

X::operator T( )

where **T** is a type name. The statement defines a conversion from a user-defined type **X** to **T**.

Example 5-5: A **complex** type is conversed to **double** and **int** types.

//---
// File: **complex.h**
// This program describes the definition of a **complex** class with type conversion.
//---
```
1    class complex{
2        double real, imag;
3    public:
4        complex(double =0, double =0);
5        operator double();        //conversion operator
6        operator int();           //conversion operator
7    };
```
//---
// File: **complex.cpp**
// The program presents the implementation detail of the **compex** class.
//---
```
1    #include "complex.h"
2
3    complex::complex(double re, double im) : real(re), imag(im) {}
4
```

```
5    complex::operator double()
6    {    return real;}
7
8    complex::operator int()
9    {    return (int)(real);}
10
```
//--
// File: **example5_5.cpp**
// This program illustrates how to converse a complex type to int and double types.
//--
```
1    #include "complex.h"
2    #include <iostream>
3    using namespace std;
4
5    void main()
6    {
7        complex a(2.1,5.6);
8        cout<<double(a)<<endl;
9        cout<<int(a)<<endl;
10   }
```
Result:
```
1    2.1
2    2
```

In Example 5-5, the statements operator double () and operator int () in Lines 5 and 6 in complex.h mean that the objects of complex type can be conversed to double and int types.

Conversion operator is different than an ordinary overloaded operator:
- it should not return a value (not even void)
- it takes no arguments.

5.7 Examples of Operator Overloading

5.7.1 A *Complex Number* Class

Class **complex** is a concrete type. In the previous sections, we present the definitions of operator functions for it respectively. A complete class design follows the guidelines from Section 4.9. In addition, users of complex arithmetic rely so heavily on those operators that the definition of **complex** brings into play most of the basic rules for operator overloading.

Constructors and Copy Constructors

To cope with assignments and initialization of complex variables with scalars, we need a

Chapter 5 Operator Overloading

conversion of a scalar (integer or floating point number) to a complex. For example:

    complex c1;   //create c1 by real =0, imag =0

    complex c2(2.3, 4.5);   // create c2 by real =2.3, imag =4.5

    complex c3 = c2;   //call copy constructor and initialize by c2

    complex c4 = 3;   //create complex (3); then call copy constructor and

                              //initialize by   complex(3)

According to the above application for class **complex**, the following constructors are defined:

    complex(double = 0, double = 0);

    complex(const complex&);

Member and Non-Member functions

In general, minimizing the number of functions can directly manipulate the representation of an object. This can be achieved by defining only operators that inherently modify the value of their first argument, such as +=, in the class itself. Operators that simply produce a new value based on the values of its arguments, such as +, are then defined outside the class and use the essential operators in their implementation. For example,

```
1    class complex{
2    public:
3         complex& operator += (complex);   //needs to access to data members
4         //......
5    private:
6         double real,   imag;
7    }
8
9    complex operator + (complex a, complex b)
10   {
11       complex temp   = a;
12       reture temp += b;                //access data members through +=
13   }
```

Given these declarations, we can write:

```
1    int main(){
2         complex x, y,z;
3         complex c1 = x + y + z;    //c1 = operator(x, operator(y, z));
4         complex c2 =x;
5         c2 +=y;                    //c2.operator +=(y);
6    }
```

Conversion for Mixed Arithmetic

To copy with mixed arithmetic, for example,

        complex c1 = 2 + x;

Object-Oriented Programming in C++

```
        complex c2 = x +2;
```
We need to define operator + to accept operands of different types. Thus, we can define the following functions.

```
1   class complex{
2   public:
3       complex& complex +=(complex a);
4       complex& complex += (double d);
5       //......
6   private:
7       double real, imag;
8   };
9   complex operator + (complex a, complex b)
10  {
11      complex temp = a;
12      return temp += b;            //calls complex::operator +=(complex)
13  }
14  complex operator   +(complex a, double d)
15  {
16      complex temp = a;
17      return temp += b;            //calls complex::operator +=(double)
18  }
19  complex operator + (double d, complex a)
20  {
21      complex temp = b;
22      return temp += a;            //calls complex::operator +=(double)
23  }
```

Now we present the definition of a *complex* class in Example 5-6.

Example 5-6: A complex number class.

//---
// File: **complex.h**
// This program describes the definition of a **complex** class.
//---

```
1   class complex {
2   public:
3   //constructors
4       complex();
5       complex(double);
6       complex(double, double);
7
8       //copy constructor
```

```
9         complex(const complex& c);
10
11    //operator functions as members
12        complex& operator = (complex&);
13        complex& operator ++ ();
14        complex& operator += (complex);
15        complex& operator += (double);
16
17    //postfix increment operator as friend function
18         friend complex operator ++ (complex&, int);
19
20    //member function
21        void print()const;
22
23    //additional member functions
24         double getReal() const;
25         double getImag() const;
26
27   private:
28        double real, imag;
29   };
30
31   //Operator functions as non-members for mixed arithmetic
32   complex operator + (complex&, complex&);
33   complex operator + (complex&, double);
34   complex operator + (double, complex&);
35   bool operator == (complex&, complex&);
```
//--
// File: **complex.cpp**
// The program gives the definitions of the functions to implement various operations of the
// **complex** class.
//--
```
1   #include "complex.h"
2   #include <iostream>
3   using namespace std;
4
5   //constructors and copy constructor
6   complex::complex():real(0), imag(0)
7   { cout<<"constructing... real=imag=0\n";} //complex(double r=0,double i=0)
8
9   complex::complex(double r):real(r), imag(0)
```

```cpp
10      { cout<<"constructing... imag=0\n";}
11
12      complex::complex(double r,double i):real(r), imag(i)
13      { cout<<"constructing.. \n";}
14
15      complex::complex(const complex& c):real(c.real), imag(c.imag)
16      { cout<<"copy constructor..\n";}
17
18      //operator functions as members
19      complex& complex::operator = (complex& a)
20      {
21              real = a.real;
22              imag = a.imag;
23              cout<<"Overloading assignment...\n";
24              return *this;
25      }
26
27      complex& complex::operator ++ ()
28      {
29              real++;
30              imag++;
31              cout<<"Overloading operator prefix ++...\n";
32              return *this;
33      }
34
35      complex& complex::operator += (complex a)
36      {
37              real += a.real;
38              imag += a.imag;
39              cout<<"Overloading operator class += class...\n";
40              return *this;
41      }
42
43      complex& complex::operator += (double a)
44      {
45              real += a;
46              cout<<"Overloading operator class += double...\n";
47              return *this;
48      }
49
```

```cpp
50  void complex::print() const
51  {
52      cout<<real<<" +"<<imag<<"i\n";
53  }
54
55  double complex::getReal() const
56  {
57      return real;
58  }
59
60  double complex::getImag() const
61  {
62      return imag;
63  }
64
65  //Operator functions as non-members for mixed arithmetic
66  complex operator + (complex& a, complex& b)
67  {
68      cout<<"Overloading operator    class + class...\n";
69      complex temp = a;
70      return temp += b;
71  }
72  complex operator + (complex& a, double b)
73  {
74      cout<<"Overloading operator    class + double...\n";
75      complex temp = a;
76      return temp += b;
77  }
78
79  complex operator + (double a, complex& b)
80  {
81      cout<<"Overloading operator    double + class...\n";
82      complex temp = b;
83      return temp += a;
84  }
85
86  //postfix increment operator as friend
87  complex operator ++ (complex& a, int)
88  {
89      cout<<"Overloading postfix ++...\n";
90      complex temp = a;
```

```
91          a.real ++;
92          a.imag ++;
93          return temp;
94     }
95
96     //relation operator(= =) as non-member
97     bool operator == (complex& a, complex& b)
98     {
99          cout<<"Overloading operator ==...\n";
100         return a.getReal() == b.getReal() && a.getImag() == b.getImag();
101    }
```
//---
// File: **example5_6.cpp**
// This program illustrates the use of the **complex** class.
//---

```
1      #include "complex.h"
2      #include <iostream>
3      using namespace std;
4
5      int main()
6      {
7           complex c1;
8           complex c2(20.1);
9           complex c3(2.3, 10.6);
10          complex c4 = c3;
11
12          c1 = ++c2;
13          c2.print();
14          c1 = c2++;
15          c2.print();
16
17          c1 += c3;
18          c1.print();
19
20          c4 = c1 + c2;
21          c4.print();
22          c4 = 3.5 + c1;
23          c4.print();
24          c4 = c1 + 5.0;
25          c4.print();
```

```
26
27          if(operator == (c3, c4))
28                  cout<<"Two objects is equal\n";
29          else
30                  cout<<"Two objects is not equal\n";
31
32          return 0;
33  }
```

Result:

```
1   constructing... real=imag=0
2   constructing... imag=0
3   constructing..
4   copy constructor..
5   Overloading operator prefix ++...
6   Overloading assignment...
7   21.1 + 1i
8   Overloading postfix ++...
9   copy constructor..
10  copy constructor..
11  Overloading assignment...
12  22.1 + 2i
13  copy constructor..
14  Overloading operator class += class...
15  24.4 + 12.6i
16  Overloading operator   class + class...
17  copy constructor..
18  copy constructor..
19  Overloading operator class += class...
20  copy constructor..
21  Overloading assignment...
22  46.5 + 14.6i
23  Overloading operator   double + class...
24  copy constructor..
25  Overloading operator class += double...
26  copy constructor..
27  Overloading assignment...
28  27.9 + 12.6i
29  Overloading operator   class + double...
30  copy constructor..
31  Overloading operator class += double...
```

32   copy constructor..
33   Overloading assignment...
34   29.4 + 12.6i
35   Overloading operator ==...
36   Two objects is not equal

5.7.2 A *String* Class

In the above way, we will build a class that handles the creation and manipulation of strings. Class *string* is now part of the C++ standard libraries.

A **C-string** is a character sequence stored as a one-dimensional *character array* and terminated with a null character ('\0', called NUL in ASCII). Character array is an array whose components are of type *char*. For example,

    char name[16];

This statement declares an array **name** of 16 components of type *char*. Because C-strings are null terminated and **name** has 16 components, the largest string that can be stored in **name** is of length 15. When a dynamic char array is created, it is expressed by a pointer, such as, char * name.

Now we discuss how to define a string class for C-string manipulation and, at the same time, to further illustrate operator overloading.

Example 5-7: A String class.

//---
// File: **myString.h**
// This program describes the definition of a **String** class.
//---

```
1    #include <iostream>
2    using namespace std;
3
4    class String{
5    //Overload stream output and input operators
6        friend ostream& operator << (ostream&, const String&);
7        friend istream& operator >> (istream&, String&);
8
9    public:
10       String(char* = " ");              //constructor with default parameter
11       String(const String&);            //copy constructor
12       ~String();                        //destructor
13
14       String& operator = (String&);     //assignment operator
15       String& operator += (String&);    //test s1 += s2
16
```

```
17      bool operator !();                      //is String empty?
18      bool operator == (String&);             //test s1 == s2
19      bool operator < (String&);              //test s1 < s2
20
21      String& operator () (int, int);         //return a substring
22      int Length() const;                     //return string length
23
24  private:
25      char* str;                              //pointer to start of string
26      int length;                             //string length
27      //addition member function
28      void setString(char*);
29  };
```

//--
// File: **myString.cpp**
// The program presents the implementation detail of the **String** class. It includes the header file
// **cassert.h** because we are using the function *assert*
//--

```
1   #include "myString.h"
2   #include <iostream>
3   #include <cassert>
4   #include <iomanip>
5   #include <cstring>
6   using namespace std;
7
8   void String::setString(char* string)
9   {
10      str = new char[length + 1];
11      assert(str != 0);                       //terminate if memory not allocated
12      strcpy(str, string);
13  }
14
15  //Constructor
16  String::String(char* sPtr)
17  {
18      cout<<"Constructing...\n";
19      length = strlen(sPtr);
20      setString(sPtr);
21  }
22
23  //Copy constructor
```

165

```cpp
24    String::String(const String& s)
25    {
26        cout<<"Copy constructor.\n";
27        length = s.length;
28        setString(s.str);
29    }
30
31    //Destructor
32    String::~String()
33    {
34        cout<<"Destructor\n";
35        delete []str;
36    }
37
38    //Overload operator =
39    String& String::operator = (String& s)
40    {
41        cout<<"Operator =...\n";
42        length = s.length;
43        if(&s != this)     {              //avoid self assignment
44            delete [] str;                //prevents memory leak
45            length = s.length;
46            setString(s.str);   }
47        else
48            cout<<"Attempted assignment of a string to itself\n";
49        return *this;
50    }
51
52    //Overload operator +=
53    String& String::operator += (String& s)
54    {
55        char* strTemp = str;
56        length += s.length;
57        str = new char[length + 1];
58        assert( str != 0);
59        strcpy(str, strTemp);
60        strcat(str, s.str);
61        delete []strTemp;
62        return *this;
63    }
64
```

```
65      //Is a string empty?
66      bool String::operator !()
67      {
68          return length == 0;
69      }
70
71      //Is this string equal to another?
72      bool String::operator == (String& s)
73      {
74          return strcmp(str, s.str) == 0;
75      }
76
77      //Is this string less than another?
78      bool String::operator < (String& s)
79      {
80          return strcmp(str, s.str) < 0;
81      }
82
83      //Return a substring beginning at index and of subLength
84      String& String::operator () (int index, int subLength)
85      {
86          //ensure index is in rang and subLength >= 0
87          assert(index >= 0 && index < length && subLength >= 0);
88
89          String* subStr = new String;    //empty string
90          assert( subStr != 0);
91
92          //determine length of substring
93          if((subLength == 0) || (index + subLength > length))
94              subStr->length = length - index + 1;
95          else
96              subStr->length = subLength;
97
98          delete subStr->str;
99
100         subStr->str = new char[subStr->length];
101         assert( subStr->str != 0);               //ensure space allocated
102
103         //copy substring into new string
104         strncpy(subStr->str, &str[index], subStr->length);
105         subStr->str[subStr->length] = '\0';      //terminate string
```

```
106
107        return *subStr;                        //return new string
108    }
109
110    //Return string length
111    int String::Length() const
112    {
113        return length;
114    }
115
116    //Overload output operator
117    ostream& operator << (ostream& output, const String& s)
118    {
119        output << s.str;
120        return output;                          //enable cascading
121    }
122
123    //Overload input operator
124    istream& operator >> (istream& input, String& s)
125    {
126        char temp[100];                         //buffer to store input
127        input>>setw(100)>>temp;
128        s = String(temp);                       //use assignment operator of String class
129        return input;                           //enable cascading
130    }
```

Note that the assignment operator must be explicitly overloaded for objects of the **String** type.

//--
// File: **example5_7.cpp**
// This program tests some of the operations of the **String** type.
//--

```
1   #include "myString.h"
2   #include <iostream>
3   using namespace std;
4
5   int main()
6   {
7       String s1("Hello "), s2("world! "), s3;
8       cout<<"s1 is "<<s1<<" s2 is "<<s2<<'\n';
9
10      if(!s3)
```

```
11          {
12                  cout<<"s3 is empty. asssignment s1 to s3\n";
13                  s3 = s1;
14                  cout<<"s3 is "<<s3<<'\n';
15          }
16          else
17                  cout<<"s3 is not empty\n";
18
19          if(s3 == s1)
20                  cout<<"s3 is equal to s1\n";
21          else
22                  cout<<"s3 is not equal to s1\n";
23
24          s1 += s2;
25          cout<<"Result of (s1 += s2) is "<<s1<<endl;
26
27          cout<<"The substring of s1 is "<<s1(0, 8)<<'\n';
28
29          if(s1 < s2)
30                  cout<<"s1 is less than s2\n";
31          else
32                  cout<<"s1 is greater than or equal to s2\n";
33
34          return 0;
35  }
```

Result:

1   Constructing...
2   Constructing...
3   Constructing...
4   s1 is Hello   s2 is world!
5   s3 is empty. asssignment s1 to s3
6   Operator =...
7   s3 is Hello
8   s3 is equal to s1
9   Result of (s1 += s2) is Hello world!
10 Constructing...
11 The substring of s1 is Hello wo
12 s1 is less than s2
13 Destructor
14 Destructor
15 Destructor

Most operator functions can be either member functions or non-member functions (or friend). To make an operator function be a member or non-member function of a class, please keep the following in mind:

- The function that overloads any of the operators (), [], ->, or = for a class must be declared as a member function.
- Suppose that an operator *op* is overloaded for a class, called **myClass**,
  — If the leftmost operand of *op* is an object of a different class (that is, not of tyep **myClass**), the function to overload the operator *op* must be a non-member (or friend) of the class **myClass**.
  —If an operator function to overload the operator *op* for **myClass** is a member of **myClass**, then when applying *op* on objects of **myClass**, the leftmost operand of *op* must be of type **myClass**.

Word Tips

allocated	*adj.*	分配的
application	*n.*	应用程序
associativity	*n.*	结合性
binary	*adj.*	二元的
certain	*adj.*	确定的
combination	*n.*	组合
compiling	*n.*	编译
complex number	*n.*	复数
components	*n.*	成分
concept	*n.*	概念
concrete	*adj.*	具体的；实际的
convert	*vt./vi*	转换
dynamic	*adj.*	动态的
effectively	*adv.*	有效地
exclusively	*adv.*	专门地
explicitly	*adv.*	明确地
extensible	*adj.*	可展开的
flexibility	*n.*	灵活性
guidelines	*n.*	指南
increment operator		增量运算符
library	*n.*	库
logically	*adv.*	理论上地
manipulate	*vt.*	操作
minimized	*adj.*	最少的
mutable	*adj.*	多变的
parameter	*n.*	参数
postfix	*n.*	后缀
precedence	*n.*	优先权
prefix	*n.*	前缀
previous	*adj.*	以前的
prototype	*n.*	原型
representation	*n.*	表示
respectively	*adv.*	分别地,各自地
scalars	*n.*	标量
section	*n.*	章节
stored	*v.*	储存
terminate	*v.*	表明线程被要求终止
terminology	*n.*	术语
trivially	*adv.*	平凡地
unary	*adj.*	一元的
undesirable	*adj.*	不合意的

Exercises

1. Mark the following statements as true or false and give reasons.

(1) C++ allows operators to be overloaded.

(2) C++ allows new operators to be created.
(3) Any operator can be overloaded.
(4) How an operator works on basic types cannot be changed by operator overloading.
(5) The associativity of an operator can be changed by operator overloading.
(6) The precedence of an operator cannot be changed by operator overloading.
(7) Overloaded assignment operators must be declared as non-member functions.
(8) Overloaded operator member functions for binary operators implicitly use this to obtain access to the left operand of the operator.
(9) A unary operator can be overloaded as a non-member function.
(10) A binary operator can be overloaded as a non-member function taking one argument.
(11) A conversion operator converts an object of one class into an object of another class.
(12) The ++ (or --) operators can be overloaded for both prefix and postfix incrementing (or decrementing).

2. Given the following program.

```
class Rectangle{
private:
    int length;
    int width;
public:
    Rectangle(int l = 0, int w = 0):length(l), width(w)
    {cout<<"constructing a Rectangle object..\n";}
    Rectangle & operator = (Rectangle& r);
    int getLength() const { return length;}
    int getWidth() const { return width; }
    void setLength(int l) { length = l; }
    void setWidth(int w) { width = w; }
    friend Rectangle operator + (Rectangle& r1, Rectangle& r2);
};
Rectangle& operator ++ (Rectangle& r);

int main()
{
    Rectangle rect1(5,6), rect2(7,8), rect3;
    rect3 = rect1 + rect2;
    cout<<"length="<<rect3.getLength()<<" width="<<rect3.getWidth()<<endl;

    ++rect3;
    cout<<"length="<<rect3.getLength()<<"width="<<rect3.getWidth()<<endl;
    return 0;
}
```

Define overloading operators =, + and ++ and write out the output of the program.

3. Write a program that defines a **Date** class to implement the following operations:
(1) Print a date;
(2) Increment the date by one month;
(3) Compare two dates by relation operators ==, <=, !=;
(4) Overload insertion (>>) and extraction (<<) operators for easy input and output.
    (class **Date** requires the user to input/output date in the form: dd/mm/yyyy)

4. Given the following program:
```
#include <iostream>
using namespace std;
class CVector {
  public:
    int x,y;
    CVector ();
    CVector (int,int);
    CVector(CVector&);
    CVector operator ++ ();
    friend  CVector& operator + (CVector& v1, CVector& v2);
    ~CVector();
};
```
Write out the definitions of all member functions and a friend function within the **CVertor** class and test them in a main function.

5. Given the declaration of the **Integer** class:
```
class Integer{
  public:
    Integer (int a = 0);
    Integer (const Integer& i);
    Integer & operator ++ ();
    friend bool operator< (Integer& i, Integer& j);
    void print() const ;
  private:
    int value;
};
```
(1) Write the definition of all member functions outside the **Integer** class.
(2) Write a *main* function to test them.

6. Define class **Counter** including overloading operator ++.

7. When C++ program is running, it will not check whether an array is out of bound automatically. Write a program that can do it by overloading operator [].

8. Modify class **TimeDemo** to implement input and output of the time. Overloading operator >> to input a time from the user. Overloading operator << to output the time to console.
9. Design a **Percent** class that represents the percentage of 10%, 80% and etc. Overload operators << and >> to implement the output and input of **Percent** objects; Overload operators = = and < to compare the relationship of two **Percent** objects.

Chapter 6
Inheritance

Do not multiply objects without necessity.
—W. Occam

Objectives
- To create classes by inheriting from existing classes
- The use of constructors and destructors in inheritance hierarchies
- Access control in the class and the differences between public, protected and private inheritance
- To understand multiple inheritance
- To avoid ambiguities with multiple inheritance
- To avoid ambiguities with virtual inheritance

6.1 Class Hierarchies

In the previous chapters, we discuss how to build a single class that is a representation of real objects. The class is represented by a collection of abstracted concepts with common characteristics. However, a concept does not exist in isolation always. It coexists with related concepts and derives much of its power from relationships with related concepts. It is a fundamental aspect of human intelligence to seek out, recognize, and create relationships among concepts.

For example, the concepts of a circle, a triangle and a rectangle are related in that they are both shapes; that is, they have the concept of a shape in common, that is, they have the operations of move, draw and calculation. However, we find their difference points of the circle, triangle and rectangle. Thus, we must explicitly define classes **Circle**, **Triangle** and **Rectangle** to have class **Shape** in common. Representing a circle, a triangle and a rectangle in a program without involving the notion of a shape would lose something essential. Each kind of shape is more specialized than its parent **Shape**. We can classify different kind of shape according to hierarchy shown in Figure 6-1. This could be represented in the world of classes with a class **Shape** from which we would derive the three other ones: **Circle, Triangle and Rectangle**.

Class **Shape** contains the members that are common for the three types of shape. **Circle Triangle** and **Rectangle** would be its derived classes, with specific features that are different from one type of polygon to the other.

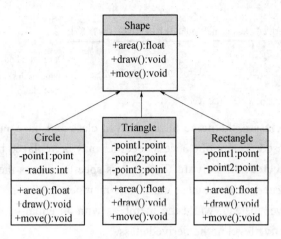

Figure 6-1　A class hierarchy

Such a set of related classes is traditionally called a ***class hierarchy***. A class hierarchy is most often a tree (Figure 6-2 (a)), but it can also be a more general graph structure (Figure 6-2 (b)).

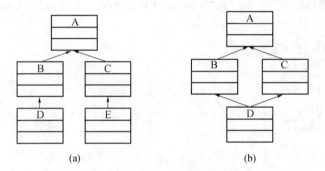

Figure 6-2　The structure of class hierarchies
(a) Tree structure; (b) Lattic structure.

A class hierarchy is created by inheritance. **Inheritance** is a key feature of C++ classes. It can allow programmers to reuse code they've already written, that is, allow to create classes which are derived from existing classes, so that they automatically include some of its "parent's" members, plus its own.

Therefore, inheritance suggests an object is able to inherit characteristics from another object. In more concrete terms, an object is able to pass on its state and behaviors to its children. In order for inheritance to work, the objects need to have characteristics in common with each other.

6.2　Derived Classes

6.2.1　Declaration of Derived Classes

In order to derive a class from another, we use a colon (:) in the declaration of the derived class using the following format:

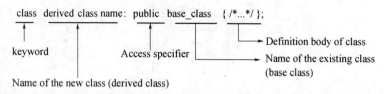

Where **derived_class** is the name of the derived class and **base_class** is the name of the class on which it is based. The public access specifier may be replaced by any one of the other access specifiers **protected** and **private**. This access specifier limits the most accessible level for the members inherited from the base class. The members with a more accessible level are inherited with this level instead, while the members with an equal or more restrictive access level keep their restrictive level in the derived class.

6.2.2 Structure of Derived Classes

For example, we consider building a program that deals with people employed by a firm. By using the data abstraction, we can define an **Employee** class. Such a program might have a class like this:

```
class Employee{
    string first Name;
    string familyName;
    Date hiringDate;
    short department;
    //...
};
```

There are some managers in the firm. We consider the manager has the properties of an employee. Thus, we might try to define a manager as follows:

```
class Manager{
    Employee emp;    //manager's employee record
    short level;
    //...
};
```

Because a manager is also an employee, the class **Employee** data is stored in the **emp** member of a **Manager** object. This may be obvious to a human reader-especially a careful reader—but there is nothing that tells the compiler and other tools that the manager is also an employee. An object **Manager** is not an **Employer**, so one cannot simply use one where the other is required.

In the real world, we do not view everything as unique; we often view something as being *like* something else but with *differences* or *additions*. Managers are *like* regular employees; however, there might be differences.

The correct approach is to explicitly state that a **Manager** is an Employee, with an inheritance relationship. So, let us redefine the **Manager** class:

    class Manager: public Employee
    {...}

Class **Manager** is derived from class **Employee**, and conversely, **Employee** is a base class for **Manager**. Class **Manager** has the members of class **Employee**, such as firstName, hiringDate, etc., in addition to its own member, such as level.

A class which adds new functionality to an existing class is said to *derive* from that original class. The original class is said to be the new class's **base class**. So a derived class is often said to inherit properties from its base, so the relationship is also called **inheritance**. A **base class** is sometimes called a **superclass** and a **derived class** is called a **subclass**.

Inheritance is a mechanism by which one class acquires the properties from the data and operations of an existing class.

Base class (superclass) — the class being inherited from.

Derived class (subclass) — the class that inherits.

This terminology, however, is confusing to people who observe that the data in a derived class object is a superset of the data of an object of its base class. A derived class is larger than its base class in the sense that it holds more data and provides more functions.

A popular and efficient implementation of the notion of derived classes has an object of the derived class represented as an object of the base class, with the information belonging specifically to the derived class added at the end, shown in Figure 6-3.

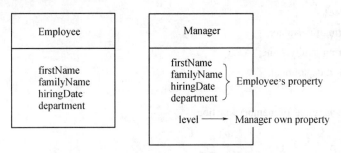

Figure 6-3   Classes **Employee** and **Manager**

Deriving class **Manager** from class **Employee** in this way makes class **Manager** be a subtype of class **Employee** so that a class **Manager** can be used wherever a class **Employee** is acceptable. A Manager is (also) an Employee, so a Manager can be used as an Employee. However, a class Employee is not necessarily a class Manager, so an Employee cannot be used as a Manager. In general, if a class Derived has a public base class Base, then a Derived can be assigned to a variable of type Base without the use of explicit type conversion. The opposite conversion, from Base to Derived, must be explicit. For example,

    void g (Manager mm, Employee ee)
    {

```
        Employee* pe = &mm;          //ok: every Manager is an Employee
        Manager* pm = &ee;           //error: not every Employee is a Manager
        pm->level=2;                 //disaster: ee doesn't have a 'level'
}
```

Notice that an object of a derived class can be treated through pointers and references. The opposite is not true.

Using a class as a base is equivalent to declaring an (unnamed) object of that class. Consequently, a class must be defined in order to be used as a base:

```
    class Employee;                  //declaration only, no definition
    class Manager: public Employee{  //error:Employee not defined
    //...
    }
```

Example 6-1: Deriving an **Manager** class from an **Employee** class.

//--
// File: **Example6_1.cpp**
// This program defines an Employee class and a Manager class.
//--

```
1    #include<iostream>
2    #include <string>
3    using namespace std;
4
5    class Employee{
6            string firstName;
7            string familyName;
8            short department;
9    public:
10           //declarations of member functions
11   };
12
13   class Manager: public Employee{
14           short level;
15   public:
16           // declarations of member functions
17   };
18   int main()
19   {
20           Employee Tom;
21           Manager Jack;
22           return 0;
23   }
```

Example 6-1 has no output because we haven't written any functions. The statements in Lines 5 to 11 define an **Employee** class, and in Lines 13 through 17 define a **Manager** class. Line 13 defines a **Manager** class that derived from **Employee** class.

6.3 Constructors and Destructors of Derived Classes

6.3.1 Constructors of Derived Classes

In principle, a derived class inherits every member of its base class except:
- its constructor and its destructor
- its operator=() members
- its friends

Derived Classes with Default Constructors

Some derived classes need constructors. If a base class has constructors, then a constructor must be invoked. Although the constructors and destructors of the base class are not inherited themselves, its default constructor (i.e., its constructor without parameters) and its destructor are always called when a new object of a derived class is created or destroyed.

The constructor of derived class is defined in the following way:

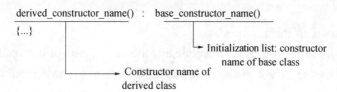

Example 6-2: Definition of a derived class with a default constructor.
//--
// File: **Example6_2.cpp**
// This program defines an Employee class and a Manager class with default constructors.
//--

```
1    #include <iostream>
2    using namespace std;
3
4    class Employee{
5    public:
6            Employee()           //default constructor
7            {
8                    cout<<"Constructor of the base class!"<<endl;
9            }
10   };
```

```
11    class Manager:public Employee{
12    public:
13         Manager() : Employee()        // initialize base class
14         {
15              cout<<"Constructor of the derived class!"<<endl;
16         }
17    };
18
19    int main()
20    {
21         Employee Tom;
22         Manager   Jack;
23         return 0;
24    }
```

Result:

1 Construction of the base class!
2 Construction of the base class!
3 Construction of the derived class!

Notice that constructor *Employee()* in Line 13 can be omitted because the constructor of the base class has no parameters.

Derived Classes' Constructors with Parameters

However, if all constructors for a base require parameters, then a constructor for that base must be explicitly called, you can specify it in each constructor definition of the derived class as follows:

derived_constructor_name (parameter list) : base_constructor_name (parameter list)
{...}

Paraineters of the base class' constructor are specified in the definition of a derived class' constructor. In this respect, the base class acts exactly like a member of the derived class. For example, the constructor of the base class **Employee** is defined as follows:

Employee::Employee(const string& n, int d) : familyName (n), department (d) {}

initializatiom list

The constructor of derived class **Manager** is defined as follows:

Manager :: Manager(const string& n, int d, int lvl) : **Employee (n, d)**   // initialize base

{
     level = lvl;   //initialize members of the derived class
}

If the constructor of the derived class directly initializs the member of its base class like

the following way,

    Manager : : Manager (const string&n, int d, int lvl)
    {
        familyName = n;   //error : family name not declared in manager
        department = d;   //error: department not declared in manager
        level = lvl;   //ok
    }

then this definition contains three errors: it fails to invoke **Employee**'s constructor, and twice it attempts to initialize members of **Employee** directly.

 A derived class constructor can specify initializers for its own members and immediate bases only; it cannot directly initialize members of its base.

Example 6-3: Definition of the derived class constructor with parameters.

//---
// File: **Example6_3.cpp**
// This program defines the constructors of an Employee class and a Manager class.
//---
```
1   #include <iostream>
2   using namespace std;
3
4   class Employee{
5           string familyName;
6           int department;
7   public:
8           Employee(const string& n, int d)
9           {
10                  familyName = n;          //initialize members
11                  department = d;
12          }
13  };
14  class Manager: public Employee{
15          int level;
16  public:
17          Manager(const string& n, int d, int lvl)
18                  :Employee(n, d)          // initialize base class
19          {
20                  level = lvl;             //initialize members
21          }
22  };
23
```

```
24    int main()
25    {
26            Employee Tom("Tom",10);
27            Manager   Jack("Jack",10,5);
28            return 0;
29    }
```

6.3.2 Destructors of Derived Classes

When a derived class object is destroyed, the program calls that object's destructor. This begins a chain (or cascade) of destructor calls in which the derived-class destructor and the destructors of the direct and indirect base classes and the classes' members execute in reverse of the order in which the constructors executed. When a derived class object's destructor is called, the destructor performs its task, and then invokes the destructor of the next base class up the hierarchy. This process repeats until the destructor of the final base class at the top of the hierarchy is called. Then the object is removed from memory.

Example 6-4: Definition of derived class constructor and destructor.

```
//--------------------------------------------------------------------
// File: Example6_4.cpp
// This program defines an Employee class and a Manager class.
//--------------------------------------------------------------------
1     #include<iostream>
2     using namespace std;
3
4     class Employee{
5         string firstName;
6         int department;
7     public:
8         Employee(const string& n, int d)
9         {
10            firstName = n;           //initialize members
11            department = d;
12        }
13        ~Employee()
14        {
15            cout<<"Destructor of the Employee class "<<firstName<<endl;
16        }
17    };
18    class Manager : public Employee{
19        int level;
20    public:
```

```
21            Manager(const string& n, int d, int lvl) : Employee(n, d)
22            {
23                level=lvl;                    //initialize members
24            }
25            ~Manager()
26            {
27                cout<<"Destructor of the Manager class"<<endl;
28            }
29       };
30
31       int main()
32       {
33            Employee Tom("Tom",10);
34            Manager    Jack("Jack",10,5);
35            Manager    Jenny("Jenny",10,2);
36            return 0;
37       }
```
Result:

1 Destructor of the Manager class

2 Destructor of the Employee class Jenny

3 Destructor of the Manager class

4 Destructor of the Employee class Jack

5 Destructor of the Employee class Tom

6.3.3 Order of Calling Class Objects

The derived class objects are constructed from the bottom up: first the base class, then the members, and then the derived class itself. They are destroyed in the opposite order: first the derived class itself, then the members, and then the base class. Members and the base class are constructed in order of declaration in the class and destroyed in the reverse order.

For example, if we have a class hierarchy such as the following:

Example 6-5: Order of calling a class object in heritance.

```
//----------------------------------------------------------------
// File: example6_5.cpp
// This program defines an Employee class and a Manager class.
//----------------------------------------------------------------
1     #include <iostream>
2     #include <string>
3     using namespace std;
4     class Date{
5          int day;
```

```
6          int month;
7          int year;
8      public:
9          Date(int d=1,int m=1,int y=2010):day(d),month(m),year(y)
10         {
11             cout<<"Date's constructor is called"<<endl;
12         }
13         ~Date()
14         {
15             cout<<"Date's destructor is called"<<endl;
16         }
17     };
18     class Employee{
19         string name;
20     public:
21         Employee():name("noname")
22         {
23             cout<<"Employee's default constructor: Employee's name is " << name << endl;
24         }
25         Employee(string na):name(na)
26         {
27             cout<<"Employee's constructor. Employee's name is "<<name <<endl;
28         }
29         ~Employee()
30         {
31             cout<<"Employee's destructor: Employee's name is "<<name <<endl;
32         }
33     };
34     class Manager:public Employee{
35         Date hiringdate;     //Date object
36         int level;
37     public:
38         Manager()
39         {
40             cout<<"Manager's default constructor is called"<<endl;
41         }
42         Manager(string na,int d,int m,int y,short lvl): **Employee(na), hiringdate(d,m,y), level(lvl)**
43         {
44             cout<<"Manager's constructor is called"<<endl;
45         }
46         ~Manager()
```

```
47              {
48                      cout<<"Manager's destructor is called"<<endl;
49              }
50      };
51      int main()
52      {
53              Manager noname;
54              Manager Jack("Jack",12,12,1991,10);
55              return 0;
56      }
```
Result:

1 Employee's default constructor:Employee's name is noname

2 Date's constructor is called

3 Manager's default constructor is called

4 Employee's constructor:Employee's name is Jack

5 Date's constructor is called

6 Manager's constructor is called

7 Manager's destructor is called

8 Date's destructor is called

9 Employee's destructor:Employee's name is Jack

10 Manager's destructor is called

11 Date's destructor is called

12 Employee's destructor:Employee's name is noname

In Line 53, a Manager object **noname** is created, causing the first three lines of output when the **Manager**'s default constructor is called. Line 38 states the **Manager**'s default constructor. The base's constructor can be omitted in the Manager's initializers because **Employee** also has default constructor. When we call the **Manager**'s default constructor, the **Employee**'s default constructor will be called implicitly although we do not call it explicitly, and then the **Date**'s default constructor will be called implicitly.

In Line 54, a Manager object **Jack** is created, causing the Lines 4 to 6 of results when the **Manager**'s constructor is called. Line 42 states the **Manager**'s constructor. When we called the **Manager**'s constructor, the **Employee**'s constructor will be called first, and then the **Date**'s constructor will be called.

Finally, the **Manager** object goes out of scope and the destructor is called. The objects are destroyed in the opposite order. Firstly the **Manager**'s destructor is called. Secondly, the **Date**'s destructor is called. Last, the **Employee**'s destructor is called.

In a word, the construction order of the **Manage** object is as follows: **Employee -> Date -> Manager**. Whereas, the destruction order of the **Manage** object is as follows: **Manager -> Date ->Employee.**

6.3.4 Inheritance and Composition

It turns out that much of the syntax and behavior are similar for both composition and inheritance (which makes sense; they are both ways of making new classes from existing classes). However, composition and inheritance are different.

Inheritance represents the ***is-a*** relationship. In the *is-a* relationship, an object of a derived class also can be treated as an object of its base class. For example, a manager is an employee, so any properties and behaviors of an employee are also properties of a manager.

By contrast, ***composition*** represents the ***has-a*** relationship. In the *has-a* relationship, an object contains one or more objects of other classes as members. In Example 6-5, a **Manager** class has **hiringdate** property (in Line 35) which is an object of class **Date**. In other word, we can see the **hiringdate** object is data member of the class **Manager**.

The UML diagram of these two kinds of class relationship is shown in Figure 6-4.

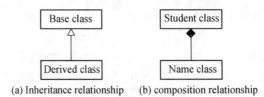

(a) Inheritance relationship   (b) composition relationship

Figure 6-4   The UML diagram of two kinds of class relationship

---

Think These Over

1. If a base class has a default constructor, does the constructor in the derived class need to explicitly call a base class constructor? If a base class has a constructor with parameters but no default values, does the derived class need to have a constructor with parameters explicitly?
2. How do we differentiate between composition and inheritance?
3. What do the advantages of inheritance and composition include?

6.4   Member Functions of Derived Classes

We need to define the member functions of classes **Employee** and **Manager**. For example,

```
class Employee{
    string firstName;
    string familyName;
public:
    void print() const;
    string full_ name() const
    { return firstName + familyName; }
```

};
class Manager: public Employee{
public:
    void print() const;
};

In this example, the *print* function of derived class **Manager** can use the public or protected members of its base **Employee** in the derived class itself. For example,

void   Manager::print() const
{
    cout<<"name is"<<full_name)()<<endl;
}

However, the *print* function of derived class **Manager** cannot use a base class' private members:

void   Manager::print() const
{
    cout<<"name is"<<familyName<<endl;   //*error*
}

This second version of *Manager::print()* will not compile. A member of a derived class has no special permission to access private members of its base class, so **familyName** is not accessible to *Manager::print()*.

Typically, the cleanest solution is for the derived class to use only the public members of its base class. For example,

void   Manager::print() const
{
    Employee::print();   //*print Employee information*
    cout<<level;   // *print Manager-specific information*
}

 The scope operator (::) must be used because function *print* has been redefined in derived class **Manager**. Such reuse of names is typical.

Overriding Member Functions

A **Manager** object has access to all member functions in class **Employee**, as well as to any member functions, such as *full_name*, that the declaration of the **Manager** class might add.

Sometimes, to keep the same function name as that of its base class, class Manager can also **override** a base class function. *Overriding a member function means changing the implementation of a base class function in a derived class.* When you make an object of the derived class, the correct function is called.

When you override a function, it must agree in return type and in signature with the

187

function in the base class. The signature is the function prototype rather than the return type: that is, the name, the parameter list, and the keyword const if used.

> When a derived class creates a function with the same return type and signature as a member function in the base class, but with a new implementation, it is said to be ***overriding*** that method. The *signature* of a function is its name, as well as the number and type of its parameters. The signature does not include the return type.

Example 6-6: Overriding the *print* function of a Manager class.

```
//-------------------------------------------------------------------
// File: Example 6_6.cpp
// This program defines an Employee class and a Manager class.
//-------------------------------------------------------------------
1    #include <iostream>
2    #include <string>
3    using namespace std;
4
5    class Employee {
6         string firstName;
7         string familyName;
8         short department;
9    public:
10        Employee(string firstname, string familyname, short depart)
11        {
12            firstName = firstname;
13            familyName = familyname;
14            department = depart;
15        }
16        void print() const
17        {
18            cout<<"name is "<<full_name()<<endl;
19            cout<<"department is" <<department<<endl;
20        }
21        string full_name() const
22        {   return firstName+" "+family_name;   }
23   };
24   class Manager: public Employee {
25        short level;
26   public:
27        Manager(string firstname, string familyname, short depart, short lvl) :
28              Employee(firstname, familyname, depart)
29        {
30            level=lvl;
```

```
31              }
32              void print() const    //overwriting the print function
33              {
34                      Employee::print();
35                      cout<<level<<endl;
36              }
37      };
38      int main()
39      {       Employee Tom("Tom","Felton", 12);
40              Manager Jack("Jack","Sparrow",15,11);
41              Tom.print();
42              Jack.print();
43              return 0;
44      }
```
Result:
```
1       name is Tom Felton
2       department is12
3       name is Jack Sparrow
4       department is15
5       11
```

In Line 32, derived class **Manager** overrides the *print* function. In Line 39, when an **Employee** object **Tom** is created, the **Employee** constructor is called. In Line 40, when a **Manager** object **Jack** is created, the **Employee** constructor and then the **Manager** constructor are called respectively.

In Line 41, the **Employee** object calls its *print* function, and the **Manager** object then calls its *print* function in Line 42. The output reflects that the correct functions were called. Finally, the two objects go out of scope and their destructors are called respectively.

If you have overridden the base function, it is still possible to call it by fully qualifying the name of the method. You do this by writing the base name, followed by two colons and then the method name. For example, *Employee::print()*.

Think It Over

Why do we need to override the member functions of the derived class?

6.5 Access Control

6.5.1 Access Control in Classes

As mentioned in Section 3.4, a member of a class can be private, protected, or public:
- If it is private, its name can be used only by member functions and friends of the class in

which it is declared.

• If it is protected, its name can be used only by member functions and friends of the class in which it is declared and by member functions and friends of classes derived from this class.

• If it is public, its name can be used by any function.

This reflects the view that there are three kinds of functions accessing a class: functions implementing the class (its friends and members), functions implementing a derived class (the derived class' friends and members), and other functions. This can be presented graphically, shown in Figure 6-5.

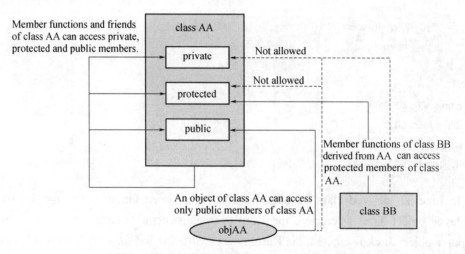

Figure 6-5   Three kinds of access control of a class

Example 6-7: Access control to the members of a **smap** class.

//--
// File: **example6_7.cpp**
// This program verifies access control to the members of a samp class.
//--

```
1    #include <iostream >
2    using namespace std;
3
4    class samp{
5        int a;
6    protected:
7        int b;
8    public:
9        int c;
10       samp(int n,int m)
11       { a=n;b=m;}
```

```
12        int geta() {return a; }
13        int getb() {return b; }
14     };
15     int main()
16     {
17        samp sa(20,30);
18        sa.b=99;                        //error
19        sa.c=50;
20        cout<<sa.geta()<<endl;
21        cout<<sa.getb()<<endl;
22        cout<<sa.c<<endl;
23        return 0;
24     }
```

This example will not compile because variable **b** is the protected member of class **samp**. It cannot be access outside the class. If we delete Line 18, we will see the output of the example as this:

1   20
2   30
3   50

6.5.2 Access to Base Classes

Like a member, a base class can be declared private, protected, or public when another class is derived from this base class. For example,

    class X : public B{/*   */};
    class Y : protected B{/*   */};
    class Z : private B{/*   */};

The keyword after the colon (:) denotes the most accessible level the members inherited from the class that follows it will have.

Public derivation makes the derived class a subtype of its base; this is the most common form of derivation. Since public is the most accessible level, by specifying this keyword the derived class will inherit all the members with the same levels they had in the base class.

Protected and **private** derivations are used to represent implementation details. **Protected** bases are useful in class hierarchies in which further derivation is the norm. **Private** bases are most useful when defining a class by restricting the interface to a base so that stronger guarantees can be provided.

The access specifier for a base class can be left out. In that case, the base defaults to a **private** base for a class and a public base for a struct.

For example,

    class XX:B{/*...*/};   //B is a private base for a class type struct YY:B{/*...*/};   //B is a public base for a struct type

For readability, it is best always to use an explicit access specifier.

For example,

    class base : private B{};   // B is a private base
    class base: protected B {};   // B is a protected base
    class base: public B{};   // B is a public base

If a member is private in the base class, then only the base class and its friends may access that member. The derived class has no access to the private members of its base class, nor can it make those members accessible to its own users.

If a base class member is public or protected, then the access label used in the derivation list determines the access level of that member in the derived class:

- In public inheritance, the members of the base retain their access levels: The public members of the base are public members of the derived and the protected members of the base are protected in the derived.
- In protected inheritance, the public and protected members of the base class are protected members in the derived class.
- In private inheritance, all the members of the base class are private in the derived class.

We can see this clearly in Table 6-1.

Table 6-1   Access control

	Base class	Derived class		General users
Private	Private	private	NA	NA
	Protected	private	A	NA
	Public	private	A	NA
Protected	Private	private	NA	NA
	Protected	protected	A	NA
	Public	protected	A	NA
Public	Private	private	NA	NA
	Protected	protected	A	NA
	Public	public	A	A

Note: A—Access; NA—No access.

Example 6-8: Levels of inheritance.

//--
// File: **example6_8.cpp**
// This program defines some classes to test levels of inheritance.
//--

```
1    class Base{
2    public:
3        int m1;
4    protected:
```

```cpp
5        int m2;
6    private:
7        int m3;
8    };
9    class Privateclass: private Base{
10   public:
11       void test()
12       {
13           m1=1;           //ok
14           m2=2;           //ok
15           m3=3;           //error
16       }
17   };
18   class DerivedFrompri: public Privateclass{
19   public:
20       void test()
21       {
22           m1=1;           //error
23           m2=2;           //error
24           m3=3;           //error
25       }
26   };
27   class Protectedclass: protected Base{
28   public:
29       void test()
30       {
31           m1=1;           //ok
32           m2=2;           //ok
33           m3=3;           //error
34       }
35   };
36   class DerivedFrompro: public Protectedclass{
37   public:
38       void test()
39       {
40           m1=1;           //ok
41           m2=2;           //ok
42           m3=3;           //error
43       }
```

```
44   };
45   class Publicclass: public Base{
46   public:
47       void test()
48       {
49               m1=1;         //ok
50               m2=2;         //ok
51               m3=3;         //error
52       }
53   };
54   class DerivedFrompub:public Publicclass{
55   public:
56       void test()
57       {
58               m1=1;         //ok
59               m2=2;         //ok
60               m3=3;         //error
61       }
62   };
63   void main()
64   {
65       Privateclass pri;
66       pri.m1=1;             //error
67       pri.m2=2;             //error
68       pri.m3=3;             //error
69       Protectedclass pro;
70       pro.m1=1;             //error
71       pro.m2=2;             //error
72       pro.m3=3;             //error
73       Publicclass pub;
74       pub.m1=1;
75       pub.m2=2;             //error
76       pub.m3=3;             //error
77   }
```

This example will not compile because the members of part of the above classes are not accessed by the *main* function. Please try to correct and give its result. The levels of inheritance in Example 6-8 is shown in Figure 6-6.

 Think It Over
What is the effect of an access specifier?

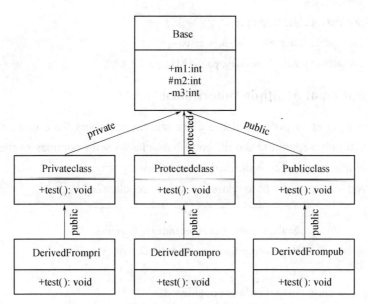

Figure 6-6 Levels of inheritance

6.6 Multiple Inheritance

In the previous sections, each class has inherited from a single parent. Such single inheritance is sufficient to describe most real-world relationships. Some classes, however, represent the blending of two classes into one.

Sometimes a class is constructed from a lattice of base classes. A derived class can itself be a base class. For example,

    class Employee{/*.*/};

    class Manager:public Employee{/*…*/}

    class Director:public Manager{/*…*/}

Because most such lattices historically have been trees, a class lattice is often called a *class hierarchy*. It can also be a more general graph structure, shown in Figure 6-7. For example,

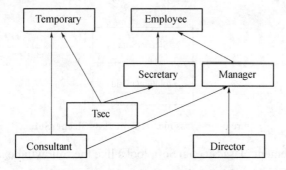

Figure 6-7 A class hierarchy

Object-Oriented Programming in C++

```
class    Temporary:{/*...*/}
class    Secretary: public Employee{/* */}
class    Tsec: public Temporary, public Secretary{/*...*/}
class    Consultant: public Temporary, public Manager{/*...*/}
```

6.6.1 Declaration of Multiple Inheritance

In C++, it is perfectly possible that a class inherits members from more than one class. This is done by simply separating the different base classes with commas in the derived class declaration. Multiple inheritance is defined in the following way:

class derived_class: public base_class1, public base_class2
{ /*...*/ };

Multiple base classes separated by commas.
Access specifiers is preceded each base class.

For example, a class sleeper sofa is considered,

class SleeperSofa : *public* Bed, *public* Sofa

As the name implies, it is a sofa and a bed (although not a very comfortable bed). Thus, the sleeper sofa should be allowed to inherit bed like properties.

> To address this situation, C++ allows a derived class to inherit from more than one base class. This is called **multiple inheritance.**

To see how multiple inheritance works, look at the sleeper sofa example. Figure 6-8 shows the inheritance graph for class **SleeperSofa**. Notice how this class inherits from classes **Sofa** and **Bed**. In this way, it inherits the properties of both.

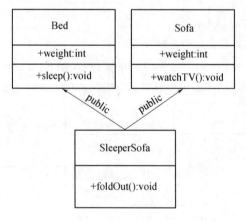

Figure 6-8   Class hierarchy of class **SleeperSofa**

The code to implement class **SleeperSofa** looks like the following example.

Example 6-9: Definition of multiple-inherited class **SleeperSofa**.

//--

```cpp
// File: example6_9.cpp
// This program defines two base classes Bed and Sofa and a multiple-inherited class SleeperSofa.
//---------------------------------------------------------------------------------------------------
1   #include <iostream>
2   using namespace std;
3   class Bed
4   {
5   public:
6       Bed(){}
7       void sleep(){ cout << "Sleep" << endl; }
8       int weight;
9   };
10  class Sofa
11  {
12  public:
13      Sofa(){}
14      void watchTV(){ cout << "Watch TV" << endl; }
15      int weight;
16  };
17  class SleeperSofa : public Bed, public Sofa
18  {
19  public:
20      SleeperSofa(){}
21      void foldOut(){ cout << "Fold out" << endl; }
22  };
23  int main()
24  {
25      SleeperSofa ss;
26      // you can watch TV on a sleeper sofa like a sofa...
27      ss.watchTV();          // Sofa::watchTV()
28      //...and then you can fold it out...
29      ss.foldOut();          // SleeperSofa::foldOut()
30      // ...and sleep on it
31      ss.sleep();
32  }
```

Class **SleeperSofa** inherits from both **Bed** and **Sofa**. This is apparent from the appearance of both classes in the class declaration. **SleeperSofa** inherits all the members of both base classes. Thus, both of the calls *ss.sleep* and *ss.watchTV* are legal. You can use a **SleeperSofa** as a **Bed** or a **Sofa**. Additionally, the class **SleeperSofa** can have members of its own, such as *foldOut*. The output of this program appears as follows:

1 Watch TV
2 Fold out
3 Sleep

6.6.2 Constructors of Multiple Inheritance

The constructing order of the multiple inheritance is similar to the single inheritance. The syntax of the constructor of derived_class is

    constructor (parameter): base_class1's constructor(parameter),
                    base_class2's constructor(parameter),
                    base_class3's constructor(parameter)

  { /*...*/ };

Example 6-10: The constructor of multiple-inherited class **SleeperSofa**.

```
//------------------------------------------------------------------------------
// File: example6_10.cpp
// This program defines two base classes Bed and Sofa and a multiple-inherited class SleeperSofa.
//------------------------------------------------------------------------------
1  #include <iostream>
2  using namespace std;
3
4  class Bed{
5  public:
6      Bed(int w): weight(w) {cout<<"Constructing Bed object...\n"; }
7      void sleep()
8      {cout<<"Sleep..\n";}
9      int weight;
10 };
11 class Sofa{
12 public:
13     Sofa(int w) : weight(w) { cout<<"Constructing Sofa object...\n";}
14     void watchTV()
15     {cout<<"Watching TV..\n"; }
16     int weight;
17 };
18 class SleeperSofa : public Bed, public Sofa{
19 public:
20     SleeperSofa(int w) : Bed(w), Sofa(w)
21     { cout<<"Constructing SleeperSofa object...\n";}
22     void foldOut()
```

```
23    {cout<<"Fold out..\n"; }
24 };
25 int main()
26 {
27    SleeperSofa ss(2);
28    ss.sleep();
29    ss.watchTV();
30    ss.foldOut();
31    return 0;
32 };
```

Result:

1 Constructing Bed object...
2 Constructing Sofa object...
3 Constructing SleeperSofa object...
4 Sleep..
5 Watching TV..
6 Fold out..

From the observation of Example 6-10, when a class has multiple base classes, base classes are initialized in a depth-first left-to-right order of appearance in the base-specifier-list. In Example 6-10, when the class object **ss** is created, the order of construction is as follows: **Bed -> Sofa -> SleeperSofa**.

6.7 Virtual Inheritance

6.7.1 Multiple Inheritance Ambiguities

Although multiple inheritance is a powerful feature, it introduces several possible problems. One is apparent in the preceding example. Notice that both **Bed** and **Sofa** contain a member **weight** in Example 6-10. This is logical because both have a measurable weight. The question is, "which **weight** does **SleeperSofa** inherit?" The answer is "both". **SleeperSofa** inherits a member **Bed::weight** and a separate member **Sofa::weight**. Because they have the same name, unqualified references to **weight** are now ambiguous. This is demonstrated in the following snippet:

```
1    #include <iostream.>
2    void fn()
3    {
4        SleeperSofa ss;
5        cout << "weight = "
6            << ss.weight        // illegal - which weight?
7            <<endl;  }
```

Object-Oriented Programming in C++

The program must now indicate one of the two weights by specifying the desired base class. The following code snippet is correct:

```
1   #include <iostream >
2   void fn()
3   {
4       SleeperSofa ss;
5       cout << "sofa weight = "
6           << ss.Sofa::weight       // specify which weight
7           << endl;
8   }
```

Although this solution corrects the problem, specifying the base class in the application function isn't desirable because it forces class information to leak outside the class into application code. In this case, function *fn* has to know that **SleeperSofa** inherits from **Sofa**. These types of so-called name collisions weren't possible with single inheritance but are a constant danger with multiple inheritance.

6.7.2   Trying to Solve Inheritance Ambiguities

In the previous case of SleeperSofa, the name collision on weight was more than a mere accident. A **SleeperSofa** doesn't have a bed weight separate from its sofa weight. The collision occurred because this class hierarchy does not completely describe the real world. Specifically, the classes have not been completely factored.

Thinking about it a little more, it becomes clear that both beds and sofas are special cases of a more fundamental concept: **furniture**. Weight is a property of all furniture. This time a latticed class hierarcly is described in Figure 6-9.

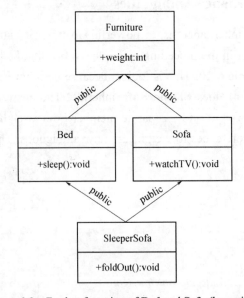

Figure 6-9   Further factoring of **Bed** and **Sofa** (by weight)

Factoring out the class **Furniture** should relieve the name collision. A C++ class hierarchy has been generated in the following program.

Example 6-11: Definition of a latticed class hirerachy.

//--
// File: **example6_11.cpp**
// This program defines four classes Furniture, Bed, Sofa, SleeperSofa.
//--

```
1   #include <iostream>
2   using namespace std;
3   // Furniture — more fundamental concept; this class has "weight" as a property
4   class Furniture
5   {
6       public:
7           Furniture(int w) : weight(w) {}
8           int weight;
9   };
10  class Bed : public Furniture
11  {
12  public:
13          Bed(int w) : Furniture(w) {}
14          void sleep(){ cout << "Sleep" << endl; }
15  };
16  class Sofa : public Furniture
17  {
18  public:
19          Sofa(int w) : Furniture(w) {}
20          void watchTV(){ cout << "Watch TV"<< endl; }
21  };
22  // SleeperSofa — is both a Bed and a Sofa
23  class SleeperSofa : public Bed, public Sofa
24  {
25  public:
26          SleeperSofa(int w) : Sofa(w), Bed(w) {}
27          void foldOut(){ cout << "Fold out"<< endl; }
28  };
29  int main()
30  {
31      SleeperSofa ss(10);
32      // the following is ambiguous; is this a Furniture::Sofa or a Furniture::Bed?
33      cout << "Weight ="<< ss.weight<< endl;      //error
34      return 0;
35  }
```

When a function or data member is called in the shared base class, another ambiguity exists. In Line 33, **weight** of the *main* function is still ambiguous. If it is not really understand why **weight** is still ambiguous, try to cast object **ss** of class **SleeperSofa** to a **Furniture** type. The explanation is straightforward. **SleeperSofa** does not inherit from Furniture directly. Both **Bed** and **Sofa** inherit from **Furniture** and then **SleeperSofa** inherits from **Bed** and **Sofa**. In memory, a **SleeperSofa** looks like Figure 6-10.

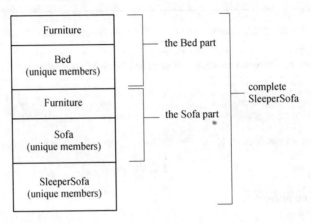

Figure 6-10   Memory layout of a SleeperSofa

You can see that a **SleeperSofa** consists of a complete **Bed** followed by a complete **Sofa** followed by some **SleeperSofa** unique stuff. Each of these sub-objects in **SleeperSofa** has its own **Furniture** part, because each inherits from **Furniture**. Thus, a **SleeperSofa** contains two **Furniture** objects!

The hierarchy shown in Figure 6-9 has not been created after all. The inheritance hierarchy created in Example 6-11 is the one shown in Figure 6-11.

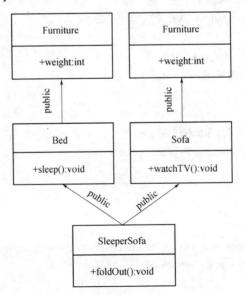

Figure 6-11   Actual result of Example 6-11

6.7.3 Virtual Base Classes

In Example 6-11, the program demonstrates this duplication of the base class. Assum that we write a main function like the following:

```
1  void main()
2  {
3       SleeperSofa ss(10);
4       //   solving ambiguous.
5       SleeperSofa* pSS = &ss;
6       Sofa* pSofa = (Sofa)*pSS;
7       Furniture* pFurniture = (Furniture)*pSofa;
8       cout << "Weight = "<< pFurniture->weight<< endl;
9  }
```

Lines 5 to 7 specifies exactly which **weight** object by recasting the pointer **SleeperSofa** first to a **Sofa*** and then to a **Furniture***. But **SleeperSofa** containing two **Furniture** objects is nonsense. **SleeperSofa** needs only one copy of **Furniture**. We want **SleeperSofa** to inherit only one copy of Furniture, and we want **Bed** and **Sofa** to share that one copy. C++ calls this **virtual inheritance** because it uses the **virtual** keyword.

Because a class can be an indirect base class to a derived class more than once, C++ provides a way to optimize the way such base classes work.

> Virtual base classes offer a way to save space and avoid ambiguities in class hierarchies that use multiple inheritance. One mechanism for specifying share a common base is a **virtual base class**.

Armed with this new knowledge, we return to class **SleeperSofa** and implement it as follows. In this example, virtual inheritance is using virtual base class, then **Bed** and **Sofa** classes can share a common base.

Example 6-12: Definition of the SleeperSofa class with virtual inheritance.

//---
// File: **example6_12.cpp**
// This program defines defines four classes Furniture, Bed, Sofa, SleeperSofa in virual inheritance.
//---

```
1   #include <iostream>
2   using namespace std;
3   // Furniture - more fundamental concept; this class has "weight" as a property
4   class Furniture
5   {
6     public:
7       Furniture(int w = 0) : weight(w) {}
8       int weight;
9   };
```

Object-Oriented Programming in C++

```
10    class Bed : virtual public Furniture
11    {
12        public:
13            Bed() {}
14            void sleep(){ cout << "Sleep" << endl; }
15    };
16    class Sofa : virtual public Furniture
17    {
18        public:
19            Sofa(){}
20            void watchTV(){ cout << "Watch TV" << endl; }
21    };
22    // SleeperSofa — is both a Bed and a Sofa
23    class SleeperSofa : public Bed, public Sofa
24    {
25        public:
26            SleeperSofa(int weight) : Furniture(weight) {}
27            void foldOut(){ cout << "Fold out" << endl; }
28    };
29    void main()
30    {
31        SleeperSofa ss(10);
32        cout << "Weight =" << ss.weight << endl;
33    }
```

Result:

1   weight=10

Notice the addition of the keyword *virtual* in the inheritance of class **Furniture** in class **Bed** and class **Sofa** in Lines 10 and 16. This says, "Give me a copy of Furniture unless you already have one somehow, in which case I'll just use that one." A **SleeperSofa** ends up looking like Figure 6-12 in memory.

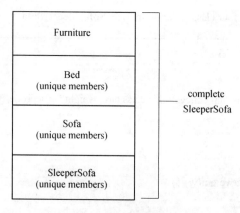

Figure 6-12   Memory layout of SleeperSofa with virtual inheritance

Here you can see that **SleeperSofa** inherits **Furniture**, and then **Bed** minus the **Furniture** part, followed by **Sofa** minus the **Furniture** part. Bringing up the rear are the members unique to **SleeperSofa**. (Note that this may not be the order of the elements in memory, but that is not important for the purpose of this discussion.)

DO use multiple inheritance when a new class needs functions and features from more than one base class. **DO** use virtual inheritance when the most derived classes must have only one instance of the shared base class. DO initialize the shared base class from the most derived class when using virtual base classes. **DON'T** use multiple inheritance when single inheritance will do.

6.7.4 Constructing Objects of Multiple Inheritance

The rules for constructing objects need to be expanded to handle multiple inheritance. The constructors are invoked in the following sequence:

First, the constructor for any virtual base classes is called in the order in which the classes are inherited.

Then the constructor for all non-virtual base classes is called in the order in which the classes are inherited.

Next, the constructor for all member objects is called in the order in which the member objects appear in the class.

Finally, the constructor for the class itself is called.

Members and objects are destroyed in the reverse order of the constructor.

Base classes are constructed in the order in which they are inherited and not in the order in which they appear on the constructor line.

Example 6-13: Definition of the SleeperSofa class with virtual inheritance.

```
//-----------------------------------------------------------------------
// File: Example6_13.cpp
// This program defines defines four classes Furniture, Bed, Sofa, SleeperSofa in virual inheritance.
//-----------------------------------------------------------------------
1    #include <iostream>
2    using namespace std;
3    class Furniture
4    {
5        int weight;
6    public:
7        Furniture(int w) :weight(w)
8        {   cout<<"Furniture's constructor is called"<<endl;   }
```

```
9        ~Furniture()
10       {   cout<<"Furniture's destructor is called"<<endl; }
11   };
12   class Bed : virtual public Furniture
13   {
14       public:
15       Bed(int w):Furniture(w)
16       {   cout<<"Bed's constructor is called"<<endl;      }
17       ~Bed()
18       {   cout<<"Bed's destructor is called"<<endl;       }
19   };
20   class Sofa : virtual public Furniture
21   {
22       public:
23       Sofa(int w):Furniture(w)
24       {   cout<<"Sofa's constructor is called"<<endl;     }
25       ~Sofa()
26       {   cout<<"Sofa's destructor is called"<<endl;      }
27   };
28   // SleeperSofa — is both a Bed and a Sofa
29   class SleeperSofa : public Bed, public Sofa
30   {
31       bool fold;    // if fold? true:fold, false: not fold
32     public:
33       SleeperSofa(int w, bool f):Furniture(w),Bed(w),Sofa(w),fold(f)
34       {   cout<<"SleeperSofa's constructor is called"<<endl;   }
35       ~SleeperSofa()
36       {   cout<<"SleeperSofa's destructor is called"<<endl;    }
37   };
38   int main()
39   {
40       SleeperSofa ss(10, true);
41       return 0;
42   }
```

Result:

```
1   Furniture's constructor is called
2   Bed's constructor is called
3   Sofa's constructor is called
4   SleeperSofa's constructor is called
5   SleeperSofa's destructor is called
```

6 Sofa's destructor is called
7 Bed's destructor is called
8 Furniture's destructor is called

 Think These Over
1. How do we avoid an inheritance ambiguity in multiple inheritance?
2. In what order is the derived object constructed in multiple inheritance?

Word Tips

a mere accident 只是意外
ambiguity *n.* 歧义，模糊
ambiguous *adj.* 不明确的
apparent *adj.* 明显的
appearance *n.* 外表
application *n.* 应用
calculate *vt./vi.* 计算
calculational *adj.* 计算的
cast *vi./vt.* 转换
classify *vt.* 分类
collision *n.* 冲突
commas *n.* 用法
composition *n.* 组合
concept *n.* 概念
concrete *adj.* 具体的
consultant *n.* 顾问
conversion *n.* 转换
demonstrate *vt.* 演示
derived *adj.* 派生的
desired *adj.* 所需的
directly *adv.* 直接地
distinction *n.* 区别
duplication *n.* 副本
element *n.* 元素
essential *adj.* 必不可少的
exactly *adv.* 准确地
explanation *n.* 解释
explicitly *adv.* 明确地
extensible *adj.* 可扩展的
feature *n.* 特征

fundamental *adj.* 基本的
further factoring 进一步分解
generate *vt.* 生成
graph *n.* 图形，图表
graphically *adv.* 图形地
hierarchy *n.* 层次，层次体系
implicitly *adv.* 隐含地
indicate *vt.* 指明
indirect *adj.* 间接的
inheritance *n.* 继承
lattice *n.* 类网格
layout *n.* 布局
leak *vt./vi.* 泄漏
legal *adj.* 合法的
logical *adj.* 合乎逻辑的
maintainable *adj.* 可维护的
measurable *adj.* 可测量的
mechanism *n.* 机制
mere *adj.* 只，仅仅
mini-payroll 小型工资
minus *n.* 减号
multiple *adj.* 多的
nonsense *adj.* 毫无意义的
notice *n.* 注意
notion *n.* 概念
optimize *vi.* 优化
overriding *v.* 重写
path *n.* 路径
plus *prep.* 加
preceding *adj.* 前面的

rear *n.* 后部	state *vt.* 陈述
recast *vt.* 重变换	straightforward *adv.* 直截了当地
relieve *vt.* 解除	sub-object *n.* 子对象
remove *vt.* 删除	subsidy *n.* 补贴
represent *vt.* 表示	sufficient *adj.* 充足的
secretary *n.* 秘书	superset *n.* 超集
separate *adj.* 单独的	temporary *adj.* 临时的
snippet *n.* 小片段	unique *adj.* 独一无二的
specify *vt.* 指定	

Exercises

1. Answer the following questions.

(1) If a base class has a constructor with argument but no default values, then derived class needn't have a constructor with arguments explicitly. *True or false*? If *false*, please give reasons.

(2) If a base class and a derived class each include a member function with the same name, which member function will be called by an object of the derived class, assuming the scope-resolution operator is not used?

(3) Assume a class Derv that is privately derived from class Base. An object of class Derv located in a main funcfion can access

    a. public members of Derv.

    b. protected members of Derv.

    c. private members of Derv.

    d. public members of Base.

    e. protected members of Base.

    f. private members of Base.

(4) A class hierarchy

    a. shows the same relationships as an organization chart.

    b. describes "has a" relationships.

    c. describes "is a kind of" relationships.

    d. shows the same relationships as a family tree.

(5) Suppose that the derived class has a member object of another class. What sequence is an object of the derived class constructed in?

(6) What is the meaning of inheritance? What is the difference between the composition and inheritance?

(7) Explain the difference between the *private* and *protected* members of a class.

(8) Suppose a base class and a derived class both define a method of the same name and a derived class object invokes the method. What method is called?

(9) Considering the following inheritance hierarchy:

    class A {

```
private:
    int x, y;
protected:
    int z;
};
class B: public A{
private:
    int a, b, c;
};
```

(a) How many data members does B have? Write out these data members.
(b) Which data members in A are *accessed* in B?
2. Fill in the blanks in each of the following:
(1) A new class inherits properties from an existing class. So the relationship is called _____ . A new class has an object of another class as its member. So the relationship is called _____ . They are all way of reusing existing software to create new software.
(2) When deriving a class from a base class with public inheritance, public members of the base class become _____ members of the derived class, and protected members of the base class become_____ members of the derived class. _____ part of the class can't be accessed by the derived class.
(3) The derived class is derived from one existing class. The existing class is called _____. The _____ members of the base class cannot be directly accessed by the derived class, but the _____ and _____ members can be accessed by the derived class. When initializing the object of a derived class, the constructor of the _____ is invoked first.
(4) To be accessed from a member function of the derived class, data or functions in the base class must be public or _____.
(5)
```
    class Employee {
    public:
        _____     //constructor of Employee class
        {
            name=theName; number=no;
        }
        void print()
        {
            cout<<name<<number<<endl;
        }
    private:
        string name;
```

                int number ;
        };
    class Manager: _____ {
            //Manager class is derived from the Employee class (public)
            int level;
    public:
            Derived(string theName, int no, int l):_____
            {
                    level=l;
            }
            void print()
            {
                    _____ //call the print function of Employee
                    cout<<level<<endl;
            }
    };

3. Find the error(s) in each of the following and explain how to correct it.
(1)
    class X{
        int a;
    public:
        X(int x) {a=x; }
        void print () {cout<<a;}
    };
    class Y: protected X{
        int b;
    public:
        Y(int x,int y) { a=x;   b=y;}
    };
    void main()
    {
            Y obj(10,2);
            obj.print();
    }
(2)
    class baseClass{
    public:
        baseClass(int a): x(a){}
        void setX(int a){ x = a;}
    private:

```cpp
        int x;
};
class derivedClass: private baseClass {
public:
    derivedClass(int a, int b)
    { y = b;}
    void setXY(int a, int b)
    { x = a; y = b; }
private:
        int y;
};
void main()
{
    derivedClass dObject(3);
    dObject.setX(50);
    dObject.setXY(50, 60);
}
```

(3)
```cpp
#include <string>
Using namespace std;
class Employee{
private:
    string name;
    int department;
public:
    Employee(const string& n, int d)
    { name = n; department = d; }
};
class Manager: public Employee{
    int level;
public:
    Manager(const string& n, int d, int lvl)
    {
        name = n;
        department = d;
        level = lvl;
    }
};
```

(4)
```cpp
class A{
```

```cpp
        int a;
public:
    A(int x) : a(x){}
    int getA() {return a; }
};
class derived1 : private A{
public:
        derived1(int x) : A(x) {}
};
class derived2 : private A{
public:
        derived2(int x) : A(x) {}
};
class B : public derived1, public derived2{
public:
    B(int x) :derived1(x), derived2(x) {}
};
int main()
{
        B bb(20);
        bb.getA();
}
```

(5)
```cpp
        #include <iostream>
        using namespace std;
        class Person{
            int ID;
        public:
            void setID(int id) {ID = id; }
            int getID() const {return ID++;}
};
class Teacher: public Person {
public:
        void printTeacher() {cout<<getID();}
};
class Student: public Person{
public:
        void printStudent() {cout<<getID(); }
    };
class TeachAssistant: public Teacher，public Student{};
```

void f()
{   TeachAssistant ta;
    ta.setID(20);
    ta.printTeacher();
    ta.printStudent();
    ta.getID(); }

4. Write the output of the following codes.

(1)
```
#include <iostream>
#include <string>
using namespace std;
class baseClass{
public:
    baseClass(string s = "", int a = 0);
    void print() const;
protected:
     int x;
private:
     string str;
};
class derivedClass: public baseClass{
public:
    derivedClass(string s = "", int a = 0, int b = 0);
    void print() const;
private:
    int y;
};
baseClass::baseClass(string s, int a)
{    str = s; x = a; }
void baseClass::print() const
{    cout<<x<<" "<<str<<endl; }
derivedClass::derivedClass(string s, int a, int b):baseClass(s,a)
{    y = b; }
void derivedClass::print() const
{
    cout<<"derived Class: "<<y<<endl;
    baseClass::print();
}
int main()
{
```

```cpp
        baseClass baseO1;
        derivedClass derivedO1;
        baseO1.print();
        derivedO1.print();
        baseClass baseO2("This is the base class", 4);
        derivedClass derivedO2("ddd", 3, 9);
        baseO2.print();
        derivedO2.print();
        return 0;
    }
```

(2)
```cpp
    #include <iostream.h>
    class X{
    public:
        X() {cout<<"X constructing...\n"; }
        ~X() {cout<<"X destroying...\n"; }
    };
    class Y : public X {
    public:
        Y() { cout<<"Y constructing...\n"; }
        ~Y() {cout<<"Y destroying...\n"; }
    };
    class Z : public Y{
    public:
        Z() { cout<<"Z constructing...\n"; }
        ~Z() {cout<<"Z destroying...\n"; }
    };
    int main(int argc, char* argv[])
    {
        Z zz;
        return 0;
    }
```

(3)
```cpp
    #include "stdafx.h"
    #include <iostream>
    using namespace std;
    class A{
        int pA;
    public:
        A(int a) { pA = a; cout<<"constructing class A"<<endl; }
```

```cpp
        int getpA() { return pA;}
};
class B: virtual public A
{
public:
        B(int a): A(a){cout<<"constructing class B"<<endl;}
        void OnB() { cout<<"B: "<<getpA()<<endl;}
};
class C: public B, virtual public A
{
public:
        C(int a):A(a),B(a) {cout<<"constructing class C"<<endl; }
        void OnC(){ cout<<"C: "<<getpA()+1<<endl; }
};
int main()
{
        C cc(5);
        cc.OnB();
        cc.OnC();
        return 0;
}
```

(4)
```cpp
#include <iostream.h>
class base{
        int a;
public:
        base(int sa)
        {    a=sa;
             cout<<"base"<<endl;   }
};
class base1:virtual public base {
        int b;
public:
        base1(int sa,int sb):base(sa)
        {    b=sb;
             cout<<"base1"<<endl;   }
};
class base2: virtual public base{
        int c;
public:
```

```
        base2(int sa, int sc):base(sa)
        {       c=sc;
                cout<<"base2"<<endl;    }
};
class derived:public base1, public base2{
    int d;
public:
    derived(int sa,int sb,int sc,int sd) : base(sa), base1(sa,sb), base2(sa,sc)
    {       d=sd;
            cout<<"derived"<<endl;      }
};
void main()
{
    derived obj(2,4,6,8);
}
```

5. The diagram of a class structure is as follows:

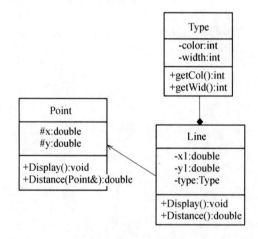

(1) Explain the relationship among classes **Point, Line** and **Type**.
(2) Explain the meaning of operators #, -, + in the class diagram.
(3) Define class **Line**. (Hint: The class only contains the declarations of all member functions.)
(4) How many data members does class **Line** have? Write out them.
(5) Define the constructor of class **Line**.
(6) Define function *Display* in class **Line**. (Hint: the function displays all information of data members in class **Line**.)
(7) Write a *main* function to test class **Line**.

6. Design and implement a mini-payroll system including three different **Person: Teacher, Student** and **TeachAssistant**.

(1) Class **Person** has data members: **number, name** and **monthly salary** and member functions to **calculate the monthly salary** and to **display** all information.

(2) Classes **Teacher** and **teachAssistant** all have another data member—**level**.

(3) The calculation of the monthly salary is defined as follows:

The **Teacher**'s monthly salary is determined by ￥50 per working hour and ￥1500 subsidy.

The **Student** gets, ￥300 subsidy per month.

The **TeachAssistant** is a part-time student and has a teacher's benefits. The salary is the student's subsidy, plus an extra salary of ￥10 per working hour.

This hierarchy can be described graphically as follows:

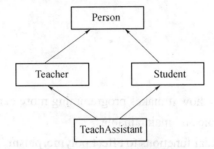

7. Derive a class called Employee2 from the Employee class in this chapter. This new class should add a type double data item called *compensation*, and also an enum type called *period* to indicate whether the employee is paid hourly, weekly, or monthly. You will derive the manager, scientist, and laborer classes from employee2 instead of employee. However, note that in many circumstances it might be more in the spirit of OOP to create a separate base class called compensation and three new classes manager2, scientist2, and laborer2, and use multiple inheritance to derive these three classes from the original manager, scientist, and laborer classes and from compensation. This way none of the original classes needs to be modified.

8. Write a program.

(1) Define a class **Vehicle** that has a constructor with parameters and protected data members *wheels* (number of wheels) and *weight* (weight of vehicle).

(2) Class **Car** is derived from **Vehicle** privately and contains data member *passenger_load* (largest number of passenger).

(3) Class **Truck** is derived from **Vehicle** privately and contains data members *passenger_load* (largest number of passenger) and *pay_load* (load capacity).

(4) Test the three classes by using a *main* function.

9. Write a program to implement the management of book and magazine sales. The requirements are as follows:

(1) Input the sales records of books and magazines;

(2) Output the names of a kind of books and magazines and their sales situation. If the situation is good, this means that the book sales is more than 500yuan monthly, and the magazine sales more than 2500 yuan; otherwise the situation is poor.

Chapter 7
Polymorphism and Virtual Functions

*General propositions do not
decide concrete cases.*
—*Oliver Wendell*

Objectives

- What polymorphism is, how it makes programming more convenient, and how it makes systems more extensible and maintainable
- To declare and use virtual functions to effect polymorphism
- The distinction between abstract and concrete classes
- To declare pure virtual functions to create abstract classes

7.1 Polymorphism

7.1.1 Concept of Polymorphism

The key idea behind OOP is polymorphism. Polymorphism is derived from a Greek word meaning "many forms". We speak of types related by inheritance as polymorphic types, because in many cases we can use the "many forms" of a derived or base type interchangeably.

In programming languages, polymprphism means that some codes or operations or objects behave differently in different contexts. For example, the + operator in C++ can implement the addition operation of different data or different types.

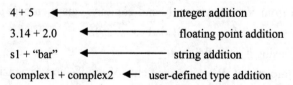

> **Polymorphism** is a key feature of object-oriented programming that allows values of different data types to be handled using a uniform interface.

The purpose of polymorphism is to implement a style of programming called *message-passing* in the literature, in which objects of various types define a common interface of operations for users.

7.1.2 Binding

When a C++ program is executed, it executes sequentially, beginning at the top of function *main*. When a function call is encountered, the point of execution jumps to the beginning of the function being called. How does the CPU know to do this?

When a program is compiled, the compiler converts each statement in your C++ program into one or more lines of machine language. Each line of machine language is given its own unique sequential address. This is no different for functions — when a function is encountered, it is converted into machine language and given the next available address. Thus, each function ends up with a unique machine language address.

> **Binding is** a process that is used to convert identifiers (such as variable and function names) into machine language addresses.

Static Binding and Dynamic Binding

By default, C++ matches a function call with the correct function definition at compile time. This is called ***static binding***. This kind of binding is also known as *compile-time binding*. For example, operator overloading and function overloading mentioned in the previous sections are that the compiler associates the operator functions and overloaded functions with their definitions in compile-time.

When the compiler matches a function call with the correct function definition at run time, this is called ***dynamic binding***. This kind of binding is also known as *run-time binding*. You declare a function with the keyword **virtual** if you want the compiler to use dynamic binding for that specific function.

Example 7-1: Static binding of member functions *f* within the classes.
```
//------------------------------------------------------------------------------
// File: example7_1.cpp
// This program illustrates how to implement static binding of member functions f within the classes.
//------------------------------------------------------------------------------
1   #include <iostream>
2   using namespace std;
3
4   class baseClass {
5   public:
6   void f() { cout << "Class baseClass" << endl; }
7   };
8
9   class derivedClass: public baseClass {
10  public:
11      void f() { cout << "Class derivedClass" << endl; }
```

```
12 };
13
14  void fn(baseClass& arg) {
15      arg.f();
16  }
17
18  int main() {
19      baseClass one;
20      fn(one);
21      derivedClass two;
22      fn(two);
23      return 0;
24  }
```
Result:
```
1  Class baseClass
2  Class baseClass
```

In Example 7-1, the *fn* function in Line 14 has a formal reference parameter of class **baseClass**. We can call the *fn* function by using an object of either class **baseClass** or class **derivedClass** as an argument. Let's look at the results generated by the statements in Lines 20 and 22. The results show only the output of **baseClass**, even though in these statements a different class object is passed as a parameter. When object two is passed as a parameter to function *fn*, we would expect that the output should be "Class derivedClass". However, actual output is "Class baseClass". Why is this? This is due to the fact that the binding of member function *f*, in the body of function *fn*, occurred at compile time. Because the formal parameter arg of function *fn* is of class **baseClass**, for statement **arg.f()** in Line 15, the compile associates function *f* of class **baseClass**.

The statement in Line 22 of Example 7-1 indicates that the actual parameter is of class **derivedClass**. Thus, when the body of function *fn* executes, logically the *f* function of object **two** should execute, which is not case. How does C++ correct this problem? To generate the output we would expect, the above example is somewhat changed in Line 6.

Example 7-2: Dynamic binding of member functions *f* within the classes.
//---
// File: **example7_2.cpp**
// This program illustrates how to implement dynamic binding of member functions *f* within the classes.
//---
```
1  #include <iostream>
2  using namespace std;
3
4  class baseClass {
```

```
5  public:
6      virtual void f() { cout << "Class baseClass" << endl; }
7  };
8
9  class derivedClass: public baseClass {
10 public:
11     void f() { cout << "Class derivedClass" << endl; }
12 };
13
14 void fn(baseClass& arg) {
15     arg.f();
16 }
17
18 int main() {
19     baseClass one;
20     fn(one);
21     derivedClass two;
22     fn(two);
23     return 0;
24 }
```

Result:

1 Class baseClass

2 Class derivedClass

The result is desirable. This is because the mechanism of virtual function is provided in Example 7-2. The binding of virtual function occurs at program execution time, not at compile time.

The two kinds of binding all have the idea of polymorphism. They are referred to as *compile-time polymorphism* and *run-time polymorphism*. Now, let's discuss the two kinds of polymorphism.

Think These Over

1. What is polymorphism?
2. What is the difference between static binding and dynamic binding?

7.2 Virtual Functions

A virtual function is a member function of the base class and is redefined by the derived class. The compiler and loader will guarantee the correct correspondence between objects and the functions applied to them. A virtual function is created using the keyword *virtual* which precedes the name of the function.

7.2.1 Definition of Vitual Functions

For example, suppose a graphics program includes several different shapes: a cube and a cuboid, and so on. Each of these classes has a member function *volume* that causes the object's volume to be calculated. We define a **Shape** class according to analysis of a class hierarchy, shown in Figure 7-1.

```
class Shape
{
 public:
    Shape();
    virtual int volume();
};
```

To enable this kind of behavior, we declare function *volume* in the base class as a **virtual function**, and we **override** *volume* in each of the derived classes to calculate the volume of appropriate shape.

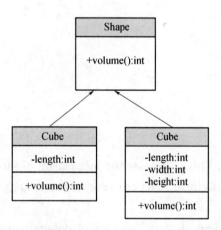

Figure 7- 1   The hierarchy of classes **Shape**, **Cube** and **Cuboid**

When a derived class inherits the class containing the virtual function, it has ability to redefine the virtual functions. A virtual function has a different functionality in the derived class. *The virtual function within the base class provides the form of the interface to the function.* Virtual function implements the philosophy of one interface and multiple methods.

Virtual functions are accessed using a base class pointer. A pointer to the base class can be created. A base class pointer can contain the address of the derived object as the derived object contains the subset of base class object. Every derived class is also a base class. When a base class pointer contains the address of the derived class object, at runtime it is decided which version of virtual function is called depending on the type of object contained by the pointer. Here is a program which illustrates the working of virtual functions.

Example 7-3: Definition of the class with virtual member functions.

//--
// File: **example7_3.cpp**
// This program illustrates the definition of a virual function within class Shape.
//--

```
1   #include<iostream>
2   using namespace std;
3   class Shape
4   {
5     public:
6         Shapc(){  }
7         virtual int volume()
8         {
9              cout << "Virtual function of the base class " << endl;
10             return 0;
11        }
12  };
13  class Cube: public Shape
14  {
15        int length;
16    public:
17        Cube(int l)
18        {
19            length = l;
20        }
21        virtual int volume()
22        {
23           cout <<"The volume of the cube is" ;
24           return length*length*length;
25        }
26  };
27  class Cuboid: public Shape
28  {
29        int length;
30        int width;
31        int height;
32    public:
33        Cuboid(int l,int w,int h)
34        {
35            length = l;
36            width = w;
37            height = h;
```

```
38          }
39              virtual int volume()
40              {
41                  cout << "The volume of the cuboid is ";
42                  return length*width*height;
43              }
44      };
45      void fn(Shape* s)
46      {
47          cout << s->volume()<<endl;
48      }
49      int main()
50      {
51          Cube c1(10);
52          Cuboid c2(15,16,20);
53          fn(&c1);
54          fn(&c2);
55          return 0;
56      }
```
Result:

1 The volume of the cube is 1000

2 The volume of the cuboid is 4800

The program has base class **Shape** and derived classes **Cube** and **Cuboid** which inherit base class **Shape**. The base class defines a virtual function *volume*.

Line 7 defines the virtual function *volume* of base class **Shape**. The keyword **virtual** precedes the function return type. In Line 21, derived class **Cube** derived from base class **Shape** redefines the virtual function *volume*. Likewise, in Line 39, derived classes **Cuboid** also redefines function *volume*. In Line 45, the statement

        void fn(Shape* s);

declares the *fn* function with the argument of a pointer **s** to base class **Shape**. In general, the fn is called a ***top-level*** function. In Lines 51 and 52, the statements declare objects **c1** and **c2** of classes **Cube** and **Cuboid**. In Line 53, the statement

        fn(&c1);

states that the address of actual parameter **c1** is passed to pointer **s** and the program then executes the statement in Line 47, the statement

        cout << s -> volume();

calls the virtual function *volume* of derived class **Cube**. The pointer **s** contains the address of object **c1** of derived class **Cube**. Likewise, the statement in Line 54 stae that the address of actual parameters **c2** is passed to pointer **s** of base class **Shape**. The pointer **s** contains the address of the object of derived class **Cuboid**.

When a virtual function is not defined by the derived class, the version of the virtual function defined by the base class is called. When a derived class that contains the virtual function is inherited by another class, the virtual function can be overloaded for the new derived class. This means that the virtual function can be inherited.

 Once a function is declared *virtual*, it remains virtual all the way down the inheritance hierarchy from that point, even if that the function is not explicitly declared virtual when a class overrides it. If the member function in the derived class is virtual, the keyword *virtual* can be omitted in the derived class. But the keyword *virtual* cannot be omitted in the base class.

7.2.2 Extensibility

With function *volume* defined as **virtual** in the base class, you can add as many new class types as you want without changing the *fn* function. In a well-designed OOP program, most or all of your functions will follow the model of function *fn* and communicate only with the base-class **interface**. Such a program is ***extensible*** because you can add new functionality by inheriting new data types from the common base class. The functions that manipulate the base class interface will not need to be changed at all to accommodate the new classes. Here is the **Shape** example with more virtual functions and a number of new classes, all of which work correctly with the old, unchanged *fn* function. The extensibility of class hierarchy mentioned in Example 7-3 is shown in Figure 7-2.

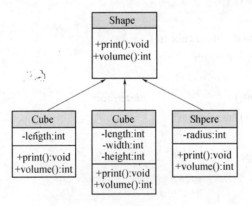

Figure 7-2 Extensibility of class Shape

Example 7-4: Extensibility of class **Shape**.

```
//---------------------------------------------------------------
// File: example7_4.cpp
// This program extends a new Sphere class without changing the fn function.
//---------------------------------------------------------------
1   #include <iostream>
```

```
2   using namespace std;
3   class Shape
4   {
5      public:
6          Shape(){  }
7          virtual int volume()              //interface
8          {
9              cout << "Virtual function of base class " << endl;
10             return(0);
11         }
12         virtual viod print()              //interface
13         {   cout<<"This is a common shape\n";   }
14  };
15
16  class Cube: public Shape
17  {
18         int length;
19     public:
20         Cube(int l)
21         {
22             length = l;
23         }
24         virtual int volume()
25         {
26           cout <<"The volume of the cube is" ;
27           return length*length*length;
28         }
29         virtual viod print()              //interface
30         {   cout<<"This is a Cube shape\n";    }
31  };
32  class Cuboid:public Shape
33  {
34         int length;
35         int width;
36         int height;
37      public:
38         Cuboid(int l,int w,int h)
39         {
40             length = l;
41             width = w;
42             height = h;
```

```cpp
43        }
44        virtual int volume()
45        {
46             cout << "The volume of the cuboid is ";
47             return length*width*height;
48        }
49        virtual viod print()              //interface
50            {   cout<<"This is a Cuboid shape\n";   }
51   };
52   class Sphere: public Shape
53   {
54        int radius;
55   public:
56        Sphere(int r)
57        {
58            radius = r;
58        }
60        virtual int volume()
61        {
62           cout <<" The volume of the sphere is " ;
63           return (int)3*3.14*radius*radius*radius/4;
64        }
65        virtual viod print()              //interface
66            {   cout<<"This is a Sphere shape\n";   }
67   };
68   void fn(Shape* s)
69   {
70       s->print();
71       cout << s->volume()<<endl;
72   }
73   int main()
74   {
75        Cube c1(10);
76        Cuboid c2(15,16,20);
77        Sphere sp(30);
78        fn(&c1);
79        fn(&c2);
80        fn(&sp);
81        return 0;
82   }
```

Result:

1 This is a Cube shape
2 The volume of the cube is 1000
3 This is a Cuboid shape
4 The volume of the cuboid is 4800
5 This is a Sphere shape
6 The volume of the sphere is 63585

7.2.3 Principle of Virtual Functions

How can *dynamic binding* happen by virtual functions? All the work goes on behind the scenes by the compiler, which installs the necessary dynamic binding mechanism when you ask it to (you ask by creating *virtual functions*). Because programmers often benefit from understanding the mechanism of virtual functions in C++, this section will elaborate on the way the compiler implements this mechanism. The keyword **virtual** tells the compiler it should not perform static binding. Instead, it should automatically install all the mechanisms necessary to perform dynamic binding. This means that if you call functions *volume* for derived class objects *through an address for the base class* **Shape**, you will get the proper function.

To accomplish this, the typical compiler creates a single virtual function table (called the VTABLE) for each class that contains **virtual** functions. The compiler places the addresses of the virtual functions for that particular class in the VTABLE. In each class with virtual functions, it secretly places a pointer, called the *vpointer* (abbreviated as VPTR), which points to the VTABLE for that object, shown in Figure 7-3. When you make a virtual function call through a base class pointer (that is, when you make a polymorphic call), the compiler quietly inserts code to fetch the VPTR and look up the function address in the VTABLE, thus calling the correct function and causing late binding to take place.

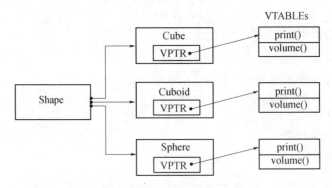

Figure 7-3   Virtual tables of Example 7-4

In Example 7-4, all of this — setting up the VTABLE for each class of **Cube**, **Cuboid** and **Shape**, initializing the VPTR, inserting the code for the virtual function call — happens automatically, so you do not have to worry about it. With virtual functions, the proper function is called for an object, even if the compiler cannot know the specific type of the object.

7.2.4 Virtual Destructors

One thing recommended for class with pointer member variables is that these class should have the destructor (see Section 4.6.2). The destructor is automatically executed when the class object goes out of the scope. Then, if the object creates dynamic objects, the destructor can be designed to deallocate the storage for them.

Base class destructors should always be virtual. Suppose you use delete with a base class pointer to a derived class object to destroy the derived class object. If the base class destructor is not virtual then delete, like a normal member function, calls the destructor for the base class, not the destructor for the derived class. This will cause only the base part of the object to be destroyed. The Example 7-5 program shows how this looks.

Example 7-5: Tests non-virtual destructors.

```
//-----------------------------------------------------------------------
// File: example7_5.cpp
// This program defines a Base class with a non-virtual destructor.
//-----------------------------------------------------------------------
1    #include <iostream>
2    using namespace std;
3
4    class Base
5    {
6    public:
7        ~Base()          //non-virtual destructor
7        { cout << "Base destroyed\n"; }
9    };
10
11   class Derived : public Base
12   {
13   public:
14       ~Derived()
15       { cout << "Derv destroyed\n"; }
16   };
17
18   int main()
19   {
20       Base* pBase = new Derived;
21       delete pBase;
22       return 0;
23   }
```

Result:

1   Base destroyed

This shows that the destructor for the **Derived** part of the object is not called. In the listing, the base class destructor is not virtual, but you can make it so by changing the definition of the Base class destructor,

```
class Base
{
  public:
      virtual ~Base()          //virtual destructor
      { cout << "Base destroyed\n"; }
};
```

Now the result is

    Derv destroyed
    Base destroyed

Now both parts of the derived class object are destroyed properly.

 If none of the destructors has anything important to do (like deallocating memory obtained with *new*), then virtual destructors aren't important. But, in general, to ensure that derived class objects are destroyed properly, you should make destructors in all base classes virtual.

7.2.5 Function Overloading and Function Overriding

We have introduced function overloading and function overriding in the previous sections. These terms are similar, and they do similar things.

When you override a member function of the class, you create a member function in a derived class with the same name as a function in the base class and the **same signature**. If we define a function as virtual in the base class, *the function with the same name and same signature* but no keyword virtual in the derived class is also virtual function because of *overriding*.

When you overload a function, you create more than one function with the same name, but with a **different signature**. If we define a function as virtual in the base class, *the function with the same name and different signature* in the derived class is **not** virtual function. We call this **overload**.

Example 7-6: Function overloading and function overriding in inheritance.

```
//-----------------------------------------------------------
// File: Example7_6.cpp
// This program illustrates the difference between function overloading and overridding.
//-----------------------------------------------------------
1   #include <iostream>
2   #include <string>
```

```cpp
3    using namespace std;
4    class Base {
5    public:
6        virtual int f() const
7        {
8            cout << "Base::f()\n";
9            return 1;
10       }
11       virtual void f(string) const {cout<<"function f has a string parameter\n" ;}
12       virtual void g() const { cout<<" Base::g()\n";}
13   };
14   class Derived1 : public Base {
15   public:
16       void g() const
17       {   cout<<"Derived1::g()\n";   }
18   };
19   class Derived2 : public Base {
20   public:
21   // Overriding a virtual function:
22       int f() const
23       {   cout << "Derived2::f()\n"; return 2;   }
24   };
25   class Derived3 : public Base {
26   public:
27   // Change return type:
28   //     void f() const     //error- overriding virtual function differs from Base::f()
29   //{ cout << "Derived3::f()\n";}
30   };
31   class Derived4 : public Base {
32   public:
33   // Overload f function by changing argument list:
34       int f(int) const
35       {   cout << "Derived4::Overloading f()\n";      return 4;   }
36       int f() const
37       {   cout << "Derived4:: Overriding f()\n";      return 4;   }
38
39   };
40   int main()
41   {
42       string s("hello");
43       int x;
```

```
44
45        cout<<"Create the Derived1 object.\n";
46        Derived1 d1;
47        x = d1.f();
48        d1.f(s);
49        d1.g();
50
51        cout<<"Create the Derived2 object.\n";
52        Derived2 d2;
53        x = d2.f();
54
55        cout<<"Create the Derived4 object.\n";
56        Derived4 d4;
57        x = d4.f(1);
58
59        Base& br = d4;   // Upcast
60        x = br.f();
61        br.f(s);      // Base version abailable
62        return 0;
63      }
```

Result:

```
1   Create the Derived1 object.
2   Base::f()
3   function f has a string parameter
4   Derived1::g()
5   Create the Derived2 object.
6   Derived2::f()
7   Create the Derived4 object.
8   Derived4:: Overloading f()
9   Derived4::Overriding f()
10  function f has a string parameter
```

The result may not be what you expect. Let us look first at class **Derived1**. There are two virtual functions *f* and a virtual function *g* defined within class **Base**. Since class **Derived1** directly inherits two functions *f* from class **Base**, the output of the statements in Lines 47 and 48 are:

Base::f()
function f has a string parameter

However, member function *g* of class **Dervied1** overrides the one of its base class. When the statement in Line 49 executes, member function *g* of **Dervied1** is called. The output is

Derived1::g()

Similarly, because virtual function *f*, without the parameter of class **Derived2**, overrides the one of its base class, the output of the statement in Line 53 is

Derived2::f()

Class **Derived4** defines another member function *f*, with different signatures from virtual functions *f* of its base class. Consequently, when the statement in Line 57 executes, object **d4** calls overloading function *f* of **Derived4** by a parameter match, rather than function *f* of class **Base**. The output is

Derived4::f()

From this point, overriding is useful when you inherit from a base class and wish to extend or modify its functionality. Even when object **d4** is casted as the base class in Line 59, it calls overridden function *f* of **Derived4** in Line 60, not the base one.

Think These Over

1. What are the advantages of using virtual functions in inheritance?
2. What is the difference between function overloading and function overriding?

7.3 Abstract Base Classes

Often in a design, you want the base class to present *only* an interface for its derived classes. That is, you don't want anyone to actually create an object of the base class, only to upcast to it so that its interface can be used. This is accomplished by making that class *abstract*.

Abstract base classes act as expressions of general concepts from which more specific classes can be derived.

It cannot be instantiated, it exists extensively for inheritance and it must be inherited. There are scenarios in which it is useful to define class that is not intended to instantiate, because such classes normally are used as base-classes in inheritance hierarchies.

This class must be inherited. This class is mostly used as a base class. You cannot create an object of an abstract class type; however, you can use pointers and references to abstract class types.

A class that contains at least one *pure virtual function* is considered an abstract class. Classes derived from the abstract class must implement the pure virtual function or they, too, are abstract classes.

A **pure virtual function** is a function which contains no definition in the base class. You declare a pure virtual function by using a pure specifier (= 0) in the declaration of a virtual member function in the class declaration.

The general form of virtual function is:

virtual return_type function_name(para_list) = 0;

where *return_type* is the type of the value returned; *function_name* is the name of the virtual function and *para_list* is the parameter list.

Each derived class should contain the definition of virtual functions. If the derived class does not define virtual function then compiler will generate an error.

> A class which contains one or more pure virtual function is called **abstract base class**.

The following program illustrates how to define a pure virtual function.

Example 7-7: Definition of an abstract base class.

//--
// File: **example7_7.cpp**
// This program defines a Shape class with a pure virtual function.
//--

```
1   #include<iostream>
2   using namespace std;
3   class Shape
4   {
5   public:
6       Shape(){    }
7       virtual int volume()=0;
8   };
9   class Cube: public Shape
10  {
11      int length;
12      public:
13      Cube(int l)
14      {
15          length = l;
16      }
17      virtual int volume()
18      {
19          cout << " Volume function of Cube " << endl;
20          return length*length*length;
21      }
22  };
23  class Cuboid:public Shape
24  {
25      int length;
26      int width;
27      int height;
28      public:
29      Cuboid(int l,int w,int h)
30      {
31          length=l;
32          width=w;
33          height=h;
```

```
34      }
35      virtual int volume()
36      {
37              cout << " Volume function of Cuboid " << endl;
38              return length*width*height;
39      }
40 };
41 void main()
42 {
43      Shape *s;
44      // Shape s1;    //error: abstract class can not have object
45      Cube c1(10);
46      Cuboid c2(15,16,20);
47      s=&c1;
48      cout << " The volume of the cube" << s->volume() << endl;
49      s=&c2;
50      cout << " The volume of the cuboid" << s->volume() << endl;
51 }
```

Result:

1 Volume function of Cube

2 The volume of the cube 1000

3 Volume function of Cuboid

4 The volume of the cuboid 4800

In Line 7 of Example 7-7, the statement

    virtual int volume()=0;

declares the pure virtual function *volume* which has no definition. In Line 44, the statement

    Shape s1;

is wrong because the objects of the abstract class **Shape** cannot be created.

The statement in Line 7 tells the compiler to reserve a slot for a function in the VTABLE, but not to put an address in that particular slot. Even if only one function in a class is declared as pure virtual, the VTABLE is incomplete.

If the VTABLE for a class is incomplete, what is the compiler supposed to do when someone tries to make an object of that class? It cannot safely create an object of an abstract class, so you get an error message from the compiler. Thus, the compiler guarantees the purity of the abstract class. By making a class abstract, you ensure that the client programmer cannot misuse it.

Pure virtual functions are helpful because they make explicit the abstractness of a class and tell both the user and the compiler how it was intended to be used.

Note that pure virtual functions prevent an abstract class from being passed into a function *by value*. Thus, it is also a way to prevent *object slicing* (which will be described shortly). By making a class abstract, you can ensure that a pointer or reference is always used during upcasting to that class.

 The objects of abstract base class cannot be created as it does not contain definition of one or more member functions except virtual functions. The pointers of abstract class can be created and the references to the abstract class can be made.

 Think These Over
1. What is the difference between abstract base classes and virtual base classes?
2. Why is the abstract class objects not declared?
3. How to understand inheritance interface and inheritance implementation?

7.4 Case Study: A Mini System

We design a mini system that includes different types of people in the university. There are five classes, that is, **Person**, **Studen**t, **Employee**, **Faculty** and **Staff,** to be defined. A class named **Person** has two derived classes named **Student** and **Employee**. Make the **Faculty** and **Staff** derived classes of **Employee**. A person has a name, address, phone number and e-mail address. A student has a class status (freshmen, sophomore, junior, or senior). An employee has an office, salary, and date-hired. Define a class named **Date** that contains the fields year, month and day. A faculty member has office hours and a rank. A staff member has a title. Define a constant virtual *toString* function in the **Person** class and override it in each class to display the class name and the person's name. The UML diagram of these five classes is described in Figure 7-4.

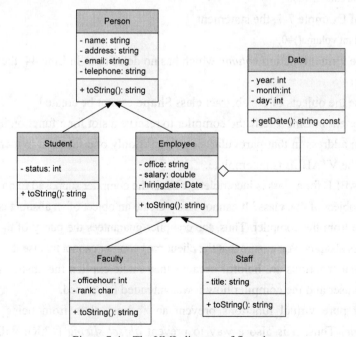

Figure 7-4   The UML diagram of five classes

Example 7-8: A mini system with five classes.
//--
// File: **people.h**
// This program defines five types of people.
//--

```
1   class Date{
2   public:
3     Date(int y, int m, int d);
4     Date(const Date& d);
5     string getDate() const;
6   private:
7     int year, month, day;
8   };
9
10  class Person{
11  public:
12    Person(string strN, string strA, string strP, string strEm);
13    virtual string toString();
14  private:
15    string name, address, phone, email;
16  };
17
18  class Student: public Person{
19  public:
20    Student(int st, string strN, string strA, string strP, string strEm);
21    string toString();
22  private:
23    int status;
24  };
25
26  class Employee: public Person{
27  public:
28    Employee(string strOf, double sal, Date date, string strN, string strA, string strP, string strEm);
29    string toString();
30  private:
31    string office;
32    double salary;
33    Date hiringDate;
34  };
35
36  class Faculty : public Employee{
```

```
37   public:
38   Faculty(int ohour, char r, string strOf, double sal, Date date, string strN, string strA, string strP, 39 string strEm);
39      string toString();
40   private:
41      int officeHour;
42      char rank;
43   };
44
45   class Staff: public Employee{
46   public:
47      Staff(string t, string strOff, double sal, Date date, string strN, string strA, string strP,
               string strEm);
48            string toString();
49   private:
50      string title;
51   };
```
//--
// File: **people.cpp**
// The program defines the member functions of all classes.
//--
```
1    #include <iostream>
2    #include <string>
3    using namespace std;
4    #include <string.h>
5    #include "people.h"
6
7    Date::Date(int y, int m, int d)
8    { year = y; month = m; day = d; }
9    Date::Date(const Date& d)
10   {
11      year = d.year;
12      month = d.month;
13      day = d.day;
14   }
15   string Date::getDate() const
16   {
17      string strDate;
18      char temp[10];
19      sprintf(temp, "%d-%d-%d", year, month, day);
20      strDate = temp;
21      return strDate;
```

```cpp
22  }
23  // definition of the person class
24  Person::Person(string strN, string strA, string strP, string strEm): name(strN), address(strA),
        phone(strP),email(strEm){}
25  string Person::toString()
26  {
27   return "Name: " + name + "; Address: " + address + "; Phone: " + phone + "; Email: " + email;
28  }
29  // definition of the student class
30    Student::Student(int st, string strN, string strA, string strP, string strEm):
          Person(strN, strA, strP, strEm), status(st){}
31  string Student::toString()
32  {
33     string tempStatus;
34     switch(status){
35     case 1:
36         tempStatus = "Freshman";
37         break;
38     case 2:
39         tempStatus = "sophomore";
40         break;
41     case 3:
42         tempStatus = "Junior";
43         break;
44     default:
45         tempStatus = "Senior";
46     }
47     cout <<"Student Information\n";
48     return Person::toString() + " I am " + tempStatus;
49  }
50  //definition of the employee class
51  Employee::Employee(string strOf, double sal, Date date, string strN, string strA, string strP,
        string strEm):    Person(strN, strA, strP, strEm), office(strOf), salary(sal), hiringDate(date){}
52  string Employee::toString()
53  {
54     char ch[20];
55     sprintf(ch, "Salary: %6.2f", salary);
56     string temp = ch;
57     return Person::toString() + temp + "; Hiring date: " + hiringDate.getDate();
58  }
59  //definition of the faculty class
```

60 Faculty::Faculty(int ohour, char r, string strOf, double sal, Date date, string strN, string strA,
        string strP, string strEm): Employee(strOf, sal, date, strN, strA, strP, strEm), officeHour(ohour),
        rank(r){}
61 string Faculty::toString()
62 {
63   char ch[20];
64   sprintf(ch, "; Office hour: %d", officeHour);
65   string temp = ch;
66   cout <<"Faculty Information\n";
67   return Employee::toString() + ch + "; Rank: " + rank;
68 }
69 //definition of the staff class
70 Staff::Staff(string t, string strOff, double sal, Date date, string strN, string strA, string strP,
        string strEm): Employee(strOff, sal, date, strN, strA, strP, strEm), title(t){}
71 string Staff::toString()
72 {
73   cout << "Staff Information\n";
74   return Employee::toString() + "; Title: " + title;
75 }

//--
// File: **example7_8.cpp**
// This program tests these classes through dynamic polymorphism.
//--

1   #include <iostream>
2   #include <string>
3   using namespace std;
4   #include "people.h"
5
6   void test(Person& p)
7   {
8   cout << p.toString() <<endl;
9   }
10 int main()
11 {
12   Date date(2015, 05, 23);
13
14   Student st1(2, "Li Ming", "Shenliao Road, Shenyang", "1800403456", "11234@qq.com");
15   test(st1);
16
17   Faculty fa(20, 'A',"210", 9000.80, date, "Mary", "Shenliao Road, Shenyang", "1860774603",

```
              "12378@qq.com");
18    test(fa);
19
20    Staff st("administrator", "301", 5800.00, date, "Zhang", "Shenliao Road, Shenyang",
              "1390000000", "11234678@qq.com");
21    test(st);
22    return 0;
23  }
```

Result:

1   Student Information
1   Name: Li Ming; Address: Shenliao Road, Shenyang;
     Phone: 1800403456; Email: 11234@qq.com
        I am sophomore
2   Faculty Information
3   Name: Mary; Address: Shenliao Road, Shenyang;
     Phone: 1860774603; Email: 12378@qq.com
     Salary: 9000.80; Hiring date: 2015-5-23; Office hour: 20; Rank: A
4   Staff Information
6   Name: Zhang; Address: Shenliao Road, Shenyang;
        Phone: 1390000000; Email: 11234678@qq.com
        Salary: 5800.00; Hiring date: 2015-5-23; Title: administrator

Word Tips

argument *n.* 参数	omit *vt.* 省略
binding *n.* 绑定	polymorphism *n.* 多态
compile *vt.* 编译	precede *vt.* 在……之前
context *n.* 上下文	pure *adj.* 纯的
convert *vi.* 转变，转化	redefine *vt.* 重新定义
denote *vt.* 表示	resolve *vi.* 解决
display *vt.* 显示	restrict *vt.* 限制
dynamic binding 动态绑定	scenario *n.* 情景
execute *vt.* 执行	sequentially *adv.* 连续地
functionality *n.* 功能性	signature *n.* 特征
illustrate *vt./vi.* 说明	slot *n.* 空位置
instantiate *vt.* 例示	static binding 静态绑定
interchangeably *adv.* 可交换地，可替换地	subset *n.* 子集
	upcase *n.* 向上
interface *n.* 接口	variable *n.* 变量
likewise *adv.* 同样地	

Exercises

1. Write the output of the following program:

```cpp
#include<iostream.h>
class Shape
{
 public:
         static int number;
    double xCoord,yCoord;
    Shape(double x,double y):xCoord(x),yCoord(y)
    {
            cout<<"Shape's constructor"<<endl;
    }
    virtual double Area() const
    {
            cout<<"Shape's area is";
            return 0.0;
    }
    ~Shape()
    {
            cout<<"Shape's destructor"<<endl;
    }
};
class Circle:public Shape
{
public:
        Circle(double x,double y):Shape(x,y)
        {
                cout<<"Circle's constructor"<<endl;
                number++;
        }
        virtual double Area() const
        {
                cout<<"Circle's area is";
                return 3.14*xCoord*xCoord;
        }
        ~Circle()
        {
                cout<<"Circle's destructor"<<endl;
        }
};
```

```cpp
class Rectangle:public Shape{
public:
    Rectangle(double x,double y):Shape(x,y)
    {
        cout<<"Rectangle's constructor"<<endl;
        number++;
    }
    virtual double Area() const
    {
        cout<<"Rectangle's area is";
        return xCoord*yCoord;
    }
    ~Rectangle()
    {
        cout<<"Rectangle's destructor"<<endl;
    }
};
void fun(const Shape &sp)
{
    cout<<sp.Area()<<endl;
}
int Shape::number=0;
int main()
{
    Circle c(2.0,5.0);
    Rectangle r(2.0,4.0);
    fun(c);
    fun(r);
    cout<<c.number<<endl;
    cout<<r.number<<endl;
    return 0;
}
```

2. Write the output of the following program.

```cpp
class A
{
public:
    A()
    {
        t();
        cout << "i from A is " << i << endl;
```

```
        }
        void t()
        {    setI(20);    }
        virtual void setI(int m)
        {    cout <<"Set a value in Class A\n ";
            i = 2 * m;    }
    protected:
        int i;
};
class B: public A
{
public:
    B()
    {
        t();
        cout << "i from B is " << i << endl;
    }
    virtual void setI(int m)
    {    cout <<"Set a value in Class B\n ";
        i = 3 * m;    }
};
int main()
{
    A* p = new B();
    return 0;
}
```

3. Given the definition of a class **Person** as follows:
```
class Person{
public:
    Person(string n, int i):name(n),id(i){}
    virtual int Time()=0; // A person spends time doing something.
    virtual void print() {cout<<name<<id<<endl;}
private:
    string name;
    int id;
};
```
Design two classes **Student** and **Teacher**. They are derived from class **Person**.
(1) Class **Student** has the properties of name, id, classNo and studyTime per week;
    Class **Teacher** has the properties of name, id, department and workTime per work;

(2) Calculate work/student time. Calculating methods are defined as follows:

The study time of a student is the class quantity multiplied by 2 (hour) per week;

The work time of a teacher is the teaching quantity multiplied 2 (hour) per week;

(3) Display the information for classes **Student** and **Teacher**.

4. Given the definition of class **Point** as follows:

```
class Point{
public:
    Point(int xx, int yy) { x = xx; y = yy; }
    virtual void print() const {cout<<x<<y<<endl;}
private:
    int x, y;
};
```

Design class **Circle** with the properties of a centre and a radius. The centre of the circle is a point. The class **Circle** must be derived from the class **Point**. The class **Circle** can implement the operations of calculating the area and printing the centre, radius and area.

Design another class **Cylinder** with the properties of a base and a height. The base is a circle. Derive this class from the class **Circle**. The class **Cylinder** can implements the operations of calculating the volume, and printing the base, height and volume.

Suppose the following code is in a main function.

```
int main()
{
    Cylinder cy(10, 20, 3, 4);
    Circle c(23, 45, 30);
    Point *p;
    p = &c;
    p->print();
    p = &cy;
    p->print();
    return 0;
}
```

5. Given the following code:

```
class Employee {
public:
    Employee(string theName, float thePayRate);
    string getName() const;
    float getPayRate() const;
    virtual float pay(float hoursWorked);
protected:
    string name;
    float payRate;
```

    };
(1) Complete the definition of class **Employee**.

   Note: employee's payment = payRate * hoursworked.

(2) Define a class **Manager** derived from **Employee**. It has the members of **Employee** and its own member **level**.

   Note: manager's payment = payRate * level * hoursWorked

(3) Define another class **Salesman** derived from **Employee**. It also has the members of **Employee**, in addition to its own member **sale.**

   Note: salesman's payment = payRate * sale * hoursWorked

(4) Test classes Manager and Salesman by using a base class pointer.

6. Here is a class hierarchy to represent various media objects:

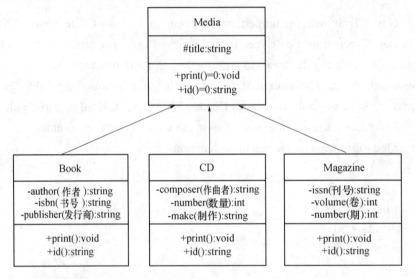

Given the definition of the **Media** base class:

```
class Media {
public:
    Media(){}
    virtual void print() = 0;
    virtual string id() = 0;
protected:
    string title;
};
```

(1) Explain the class relationship described in the hierarchy diagram.

(2) Define the **Book** and **Magazine** classes according to the hierarchy diagram. (Hint: The *print* function of each class displays all information of data members. The *id* function of each class returns a string identifier that can indicate the media feature. This identifier consists of data members of each class.)

(3) Write a *main* function to test the **Book** and **Magazine** classes. (Hint: you must define a

pointer to a **Media** object in the *main* function.)

7. Designs a class named **person** and its derived classes named **student** and **employee**.

   Assume that the declaration of the **person** class is as follows:

   ```
   class person{
   public:
           person(string strN, string strA): name(strN), address(strA){}
       virtual string toString() const;
   private:
       string name, address;
   };
   ```

   (1) Define the *toString* function in the **person** class. The returning value of the *toString* function is used to display the class information in the following format:
   Name: XXXX; Address: XXXXXXX

   (2) Define the **student** class derived from the **person** class. A student has a class status (freshmen, sophomore, junior, or senior). Override the *toString* function in the **student** class.

   (3) Define the **empolyee** class derived from the **person** class. An employee has an office and salary. Override the *toString* function in the **employee** class.

   (4) Write a **top-level** function with a parameter of **person** type and write a *main* function to test the **student** and **employee** classes.

8. Write a program.

(1) Abstract a base class **Shape**.

(2) Classes **Triangle**, **Square** and **Circle** are derived from **Shape**.

(3) Calculate the areas and perimeters of Triangle, Square and Circle.

9. Design a class hierarchy as follows:

(1) A base class **Shape** with virtual function *print*.

(2) Classes **TwoDShape** and **ThreeDShape** derived frome **Shape**. **TwoDShape** has virtual functions *area* and *perimeter*. **ThreeDShape** has virtual function *volume*.

(3) Classes **Triangle**, **Square** and **Circle** are derived from **TwoDShape**.

(4) Classes **Cube**, **Cuboid** and **Sphere** are derived from **ThreeDShape**.

   Override the definition of the print function of each derived class. Test the hierarchy by using a main function and a top function with a pointer to class **Shape**.

Chapter 8
Templates

Your quote here.
—*B. Stroustrup*

Objectives
- To use function templates to conveniently create a group of related (overloaded) functions
- To distinguish between function templates and templates functions
- To use class templates to create a group of related types
- To distinguish between class templates and template classes

8.1 Template Mechanism

Many C++ programs use common data structures like stacks, queues and lists. A program may require a queue of customers and a queue of messages. One could easily implement a queue of customers, then take the existing code and implement a queue of messages. The program grows, and now there is a need for a queue of orders. So just take the queue of messages and convert that to a queue of orders (copy, paste, find, replace). Need to make some changes to the queue implementation? Not a very easy task, since the code has been duplicated in many places. Reinventing source code is not an intelligent approach in an object oriented environment which encourages reusability. It seems to make more sense to implement a queue that can contain any arbitrary type rather than duplicating code. How does one do that? The answer is to use type parameterization, more commonly referred to as templates.

To avoid rewriting code that would be identical except for different types. Sometimes you do not simply rely on implicit type conversion or promotion match, and you cannot stuff everything into a class hierarchy.

Thus C++ templates allow one to implement a generic Queue<T> template that has a type parameter **T**. T can be replaced with actual types, for example,

> Queue<Customers>,

and C++ will generate the class Queue<Customers>. Changing the implementation of the **Queue** becomes relatively simple. Once the changes are implemented in the template Queue<T>, they are immediately reflected in the classes Queue<Customers>, Queue<Messages>, and Queue<Orders>.

Templates are very useful when implementing generic constructs like vectors, stacks, lists, queues which can be used with any arbitrary type. C++ templates provide a way to re-use source code as opposed to inheritance and composition that provide a way to re-use object code.

> Templates provide direct support for generic programming, that is, programming using types as parameters. The C++ **template** mechanism allows a type to be a parameter in the definition of a class or a function.

C++ provides two kinds of templates: **class templates** and **function templates**. Use function templates to write generic functions that can be used with arbitrary types. For example, one can write searching and sorting routines that can be used with any arbitrary type. The generic algorithms in the Standard Template Library (i.e. STL) have been implemented as function templates, and the containers have been implemented as class templates.

Inheritance and composition provide a way to reuse *object* code. The *template* feature in C++ provides a way to reuse *source* code.

8.2 Function Templates and Template Functions

8.2.1 Why We Use Function Templates?

Let's imagine that we want to write a function to compare two values and indicate whether the first is less than, equal to, or greater than the second. In practice, we would want to define several such functions, each of which could compare values of a given type. Our first attempt might be to define several overloaded functions:

```
int compare(const string &v1, const string &v2)
{
    if (v1 < v2) return -1;
    if (v2 < v1) return 1;
    return 0;
}
int compare(const double &v1, const double &v2)
{
    if (v1 < v2) return -1;
    if (v2 < v1) return 1;
    return 0;
}
```

These functions are nearly identical. The only difference between them is the type of their parameters. The function body is the same in each function. Having to repeat the body of the function for each type that we compare is tedious and error-prone. More importantly, we need to know in advance all the types that we might ever want to compare. This strategy cannot work if we want to be able to use the function on types that we don't know about. Therefore, we think that compare to be defined as a generic function.

8.2.2 Definition of Function Templates

Rather than defining a new function for each type, we can define a single *function template*. A function template is a type-independent function that is used as a formula for generating a type-specific version of the function.

> **Function templates** are special functions that can operate with generic types.

This allows us to create a function template whose functionality can be adapted to more than one type or class without repeating the entire code for each type.

 In C++, function template can be achieved using template parameters.

A template parameter is a special kind of parameter that can be used to pass a type as argument: just like regular function parameters can be used to pass values to a function, template parameters allow to pass also types to a function. These function templates can use these parameters as if they were any other regular type.

The format for declaring function templates with type parameters is:

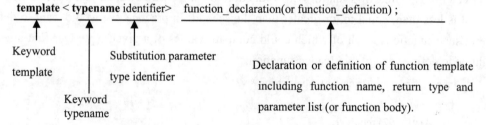

Sometimes we can define a function template in the following way:

template <class identifier> function_declaration;

The only difference between both prototypes is the use of either the keyword **class** or the keyword **typename**. Its use is indistinct, since both expressions have exactly the same meaning and behave exactly the same way.

For example, to define a function template that returns the greater one of two objects we could use:

```
1    template <typename T>    // you can also write: template<class T>
2    T getMax (T a, T b)
3    {
4        return (a>b?a:b);
5    }
```

Here we have created a function template with **T** as its template parameter. This template parameter represents a type that has not yet been specified, but that can be used in the function template as if it were a regular type. As you can see, the function template **getMax** returns the greater of two parameters of this still-undefined type.

8.2.3 Function Template Instantiation

To use this function template we use the following format for the function call:
   function_name <type> (parameters);
In last codes, to call **getMax** to compare two integer values of type *int* we can write:
   int x,y;
   getMax <int> (x,y);

When the compiler encounters this call to a template function, it uses the template to automatically generate a function replacing each appearance of **T** by the type passed as the actual template parameter (**int** in this case) and then calls it. This process is automatically performed by the compiler and is invisible to the programmer.

> The template instantiation is also called **template function**.

Let's make a complete example according to the definition of function template **getMax**.
Example 8-1: A function template and its template functions.
//--
// File: example8_1.cpp
// This program defines a funciton template getMax and its template functions.
//--

```
1    #include <iostream>
2    using namespace std;
3
4    template <class T>
5    T getMax (T a, T b)          //function template
6    {
7        T result;
8        result = (a>b)? a : b;
9        return (result);
10   }
11   int main ()
12   {
13       int i=5, j=6, k;
14       long l=10, m=5, n;
15       k = getMax<int> (i,j);    //template function
16       n = getMax<long> (l,m);   //template function
17       cout << k << endl;
```

Object-Oriented Programming in C++

```
18      cout << n << endl;
19      return 0;
20    }
```

Result:

1 6
2 10

In this case, we have used **T** as the template parameter name instead of because it is shorter and in fact is a very common template parameter name. But you can use any identifier you like.

In Example 8-1, the function template **getMax** is called twice. In Line 15, the first time with arguments of type **int**, and in Line 16, the second one with arguments of type **long**. The compiler has instantiated and then called each time the appropriate version of the function.

As you can see, the type **T** is used within the **getMax** function template even to declare new objects of that type. In Line 7, the statement

   T result;

denotes that **result** will be an object of the same type as the parameters **a** and **b** when the function template is instantiated with a specific type.

In this specific case where the generic type **T** is used as a parameter for **getMax** the compiler can find out automatically which data type has to instantiate without having to explicitly specify it within angle brackets (like we have done before specifying **<int>** and **<long>**). So we could have written instead:

   int i,j;

   getMax<int>(i,j)

Since both **i** and **j** are of type int, and the compiler can automatically find out that the template parameter can only be int. This implicit method produces exactly the same result.

Example 8-2: A function template and its template functions without explicitly type specify.

//--

// File: **example8_2.cpp**

// This program defines a funciton template getMax and its template functions.

//--

```
1    #include <iostream>
2    using namespace std;
3
4    template <class T>
5    T getMax (T a, T b) {
6        return (a>b?a:b);
7    }
8    int main ()
9    {
```

252

```
10      int i=5, j=6, k;
11      long l=10, m=5, n;
12      k = GetMax(i,j);
13      n = GetMax(l,m);
14      cout << k << endl;
15      cout << n << endl;
16      return 0;
17  }
```
Result:
```
1   6
2   10
```

Notice in this case, we called our function template **getMax** without explicitly specifying the type between *angle-brackets* <> in Lines 12 and 13. The compiler automatically determines what type is needed on each call.

8.2.4 Function Template with Different Parameter Types

Because our function template in Example 8-1 includes only one template parameter (class **T**) and the function template itself accepts two parameters, both of the same **T** type, we cannot call our function template with two objects of different types as arguments. For example,

    int i;
    long l;
    k = getMin (i, l);

This would not be correct, since our **getMax** function template expects two arguments of the same type, and in this call to it we use objects of two different types.

We can also define function templates that accept more than one type parameter, simply by specifying more template parameters between the angle brackets. For example,

    template <class T1, class T2>
    T1 getMin (T1 a, T2 b)
    {
        return (a<b? a: (T1)b);
    }

In this case, our function template **getMin()** accepts two parameters of different types and returns an object of the same type as the first parameter (T1) that is passed. For example, after that declaration we could call *GetMin* with the following types:

    int i,j;
    long l;
    i = getMin<int, long> (j, l);

or simply:

i = getMin (j, l);

even though **j** and **l** have different types, since the compiler can determine the appropriate instantiation anyway.

8.2.5 Function Template Overloading

Function templates and overloading are intimately related. One can declare several function templates with the same name and even declare a combination of function templates and ordinary functions with the same name. When an overloaded function is called, overload resolution is necessary to find the right function or template function to invoke.

For example,

```
template<class T>
T sqrt(T);
template<class T>
complex<T>   sqrt(complex<T>);
double sqrt(double);
void f(complex<double> z)
{
    sqrt(2);         //sqrt<int>(int)
    sqrt( 2 .0);     //sqrt(double)
    sqrt( z );       //sqrt<double>(complex<double>)
}
```

A function template may be overloaded in several ways. We can provide other function templates that specify the same function name but different function parameters. A function template also can be overloaded by providing non-template functions with the same function name but different function arguments.

The compiler performs a matching process to determine what function to call when a function is invoked. First, the compiler finds all function templates that match the function named in the function call and creates specializations based on the arguments in the function call. Then, the compiler finds all the ordinary functions that match the function named in the function call. If one of the ordinary functions or function template specializations is the best match for the function call, that ordinary function or specialization is used. If an ordinary function and a specialization are equally good matches for the function call, then the ordinary function is used. Otherwise, if there are multiple matches for the function call, the compiler considers the call to be ambiguous and the compiler generates an error message.

For example,

```
template<class T>
T max(T, T);
```

```
    const int s=7;
    void k()
    {
        max(1,2);           //max<int>(1, 2)
        max('a','b');       //max<char>('a', 'b')
        max ( 2.7, 4 .9);   //max<double>(2.7,4.9)
        max (s, 7);         //max<int>(int(s),7) (trivial conversion used)
        max('a',1);         //error: ambiguous (no standard conversion)
        max(2.7,4);         //error: ambiguous (no standard conversion)
    }
```

We could resolve the two ambiguities either by explicit qualification:

```
    void f()
    {
        max<int>('a',1);        //max<int>(int('a'),1)
        max<double> (2 .7, 4 ); //max<double>(2.7,double(4))
    }
```

or by adding suitable declarations:

```
    int max(int i, int j){    return max<int>(i,j);   }
    double max(int i, double d){    return max<double> ( i, d);   }
    double max (double d, int i){    return max<double> (d, i);   }
    double max(double d1, double d2){    return max<double> (d1,d2);   }
    void g()
    {
        max('a';1)     //max(int('a'),1)
        max(2 .7;4)    //max(2.7,double(4))
    }
```

8.3 Class Templates and Template Classes

8.3.1 Definition of Class Templates

Just as we can define function templates, we can also define class templates. A class template definition looks like a regular class definition, except it is prefixed by the keyword template.

Class templates are called **parameterized types**, because they require one or more type parameters to specify how to customize a "generic class" template to form a class-template specialization.

For example, here is the definition of a class template for a **Stack** class. It is possible to understand the concept of a "stack" (a data structure into which we insert items at the top and retrieve those items in last-in, first-out order) independent of the type of the items being placed in the stack. However, to instantiate a stack, a data type must be specified. This creates a wonderful opportunity for software reusability. We need the means for describing the notion of a stack generically and instantiating classes that are type-specific versions of this generic stack class. C++ provides this capability through class templates.

Creating Class Template Stack< T >

Note the **Stack** class-template definition in Example 8-3(a). It looks like a conventional class definition, except that it is preceded by the header (in Line 3)

> template< class T >

to specify a class-template definition with type parameter T which acts as a placeholder for the type of the **Stack** class to be created. The programmer need not specifically use identifier T any valid identifier can be used. The type of element to be stored on this **Stack** is mentioned generically as **T** throughout the **Stack** class header and member function definitions. Now we show how **T** becomes associated with a specific type, such as *double* or *int*. Due to the way this class template is designed, there are two constraints for non-fundamental data types used with this **Stack**, they must have a default constructor (for use in Line 40 to create the array that stores the stack elements), and they must support the assignment operator (Lines 52 and 66).

Example 8-3(a) : A stack class with parameterized types.

```
//-------------------------------------------------------------------
// File: stack.h
// This program defines the class template of a stack structure.
//-------------------------------------------------------------------
1   #include<iostream >
2   using namespace std;
3   template< class T >
4   class Stack
5   {
6   public:
7       Stack( int = 10 );      // default constructor (Stack size 10)
8       // destructor
9       ~Stack()
10      {
11          delete [] stackPtr;  // delete internal space for Stack
12      } // end ~Stack destructor
13
14      bool push( const T& ); // push an element onto the Stack
```

```cpp
15      bool pop( T& );         // pop an element off the Stack
16
17      // determine whether Stack is empty
18      bool isEmpty() const
19      {
20          return top == -1;
21      }                       // end function isEmpty
22
23      // determine whether Stack is full
24      bool isFull() const
25      {
26          return top == size - 1;
27      }                       // end function isFull
28
29   private:
30      int size;               // elements in the Stack
31      int top;                // location of the top element (-1 means empty)
32      T *stackPtr;            // pointer to internal representation of the Stack
33   };                         // end class template Stack
34
35   // constructor template
36   template< class T >
37   Stack< T >::Stack( int s )
38      : size( s > 0 ? s : 10 ),   // validate size
39        top( -1 ),                // Stack initially empty
40        stackPtr( new T[ size ] ) // allocate memory for elements
41   {
42                                  // empty body
43   }                               // end Stack constructor template
44
45   // push element onto Stack;
46   // if successful, return true; otherwise, return false
47   template< class T >
48   bool Stack< T >::push( const T &pushValue )
49   {
50      if ( !isFull() )
51      {
52          stackPtr[ ++top ] = pushValue; // place item on Stack
53          return true;                   // push successful
```

```
54        }                             // end if
55
56        return false;                 // push unsuccessful
57    }                                 // end function template push
58
59    // pop element off Stack;
60    // if successful, return true; otherwise, return false
61    template< class T >
62    bool Stack< T >::pop( T &popValue )
63    {
64        if ( !isEmpty() )
65        {
66            popValue = stackPtr[ top-- ];    // remove item from Stack
67            return true;                     // pop successful
68        }                                    // end if
69
70        return false;                        // pop unsuccessful
71    }                                        // end function template pop
```

The member function definitions of a class template are function templates. The member function definitions that appear outside the class template definition each begin with the header (in Lines 36, 47 and 61)

        template< class T >

Thus, each definition resembles a conventional function definition, except that the **Stack** element type always is listed generically as type parameter **T**. The binary scope resolution operator is used with the class-template name **Stack< T >** (in Lines 37, 48 and 62) to tie each member-function definition to the class template's scope. In this case, the generic class name is **Stack< T >**. When **doubleStack** (see Line 8 of Example 8-3(b)) is instantiated as type **Stack<double>**, the specialization of the **Stack** constructor function template uses new to create an array of elements of type **double** to represent the stack (in Line 40). The statement

        stackPtr = new T[size];

in the **Stack** class template definition is generated by the compiler in the class template specialization **Stack< double >** as

        stackPtr = new double[size];

8.3.2   Class Template Instantiation

Creating a Driver to Test Class Template Stack< T >

    Using a class template is easy. Create the required classes by plugging in the actual type for the type parameters. This process is commonly known as "*Instantiating a class*". Here is a sample driver class that uses class template **Stack**.

Chapter 8   Templates

> The process of generating a class declaration from a template class and a template argument is often called **template instantiation**. We also call it **template class**.

Now, let us consider the user program that exercises class template **Stack**. The program begins by instantiating object **doubleStack** of size 5 (in Line 8). This object is declared to be of class **Stack< double >** (pronounced "Stack of double"). The compiler associates type **double** with type parameter **T** in the class template to produce the source code for a **Stack** class of type **double**. Although templates offer software-reusability benefits, remember that multiple class-template specializations are instantiated in a program (at compile time), even though the template is written only once.

Example 8-3(b) : Test of class template **Stack**.

```
//----------------------------------------------------------------
// File: example8_3.cpp
// The program tests class template Stack in a main function.
//----------------------------------------------------------------
1   #include <iostream >
2   using namespace std;
3   #include "Stack.h"          // Stack class template definition
4
5   int main()
6   {
7      Stack< double > doubleStack( 5 );   // size 5
8
9      double doubleValue = 1.1;
10
11     cout << "Pushing elements onto doubleStack\n";
12
13     // push 5 doubles onto a doubleStack
14     while ( doubleStack.push( doubleValue ) )
15     {
16        cout << doubleValue << ' ';
17        doubleValue += 1.1;
18     }                         // end while
19
20     cout << "\nStack is full. Cannot push " << doubleValue
21        << "\n\nPopping elements from doubleStack\n";
22
23     // pop elements from doubleStack
24     while ( doubleStack.pop( doubleValue ) )
25        cout << doubleValue << ' ';
26
```

```cpp
27      cout << "\nStack is empty. Cannot pop\n";
28
29      Stack< int > intStack;        // default size 10
30      int intValue = 1;
31      cout << "\nPushing elements onto intStack\n";
32
33      // push 10 integers onto intStack
34      while ( intStack.push( intValue ) )
35      {
36          cout << intValue << ' ';
37          intValue++;
38      } // end while
39
40      cout << "\nStack is full. Cannot push " << intValue
41          << "\n\nPopping elements from intStack\n";
42
43      // pop elements from intStack
44      while ( intStack.pop( intValue ) )
45          cout << intValue << ' ';
46
47      cout << "\nStack is empty. Cannot pop" << endl;
48      return 0;
49  } // end main
```

Result:

```
1   Pushing elements onto doubleStack
2   1.1 2.2 3.3 4.4 5.5
3   Stack is full. Cannot push 6.6
4
5   Popping elements from doubleStack
6   5.5 4.4 3.3 2.2 1.1
7   Stack is empty. Cannot pop
8
9   Pushing elements onto intStack
10  1 2 3 4 5 6 7 8 9 10
11  Stack is full. Cannot push 11
12
13  Popping elements from intStack
14  9 8 7 6 5 4 3 2 1
15  Stack is empty. Cannot pop
```

Line 14 invokes function *push* to place the double values 1.1, 2.2, 3.3, 4.4 and 5.5 onto

doubleStack. The while loop terminates when the program attempts to push a sixth value onto **doubleStack** (which is full, because it holds a maximum of five elements). Note that function push returns false when it is unable to push a value onto the stack. Class **Stack** provides the function *isFull*, which the programmer can use to determine whether the stack is full before attempting a push operation. This would avoid the potential error of pushing onto a full stack.

Lines 24 and 25 invoke function *pop* in a while loop to remove the five values from the stack (in Example 8-3(b), that the values do pop off in last-in, first-out order). When the program attempts to pop a sixth value, the **doubleStack** is empty, so the **pop** loop terminates. Line 29 instantiates integer stack **intStack** with the declaration.

Stack< int > intStack;

Because no size is specified, the size defaults to 10 as specified in the default constructor (in Line 7 of Example 8-3(a)). Lines 34-38 loop and invoke *push* to place values onto **intStack** until it is full, then Lines 44 and 45 loop and invoke *pop* to remove values from **intStack** until it is empty. Once again, notice in the output that the values pop off in last-in, first-out order.

8.4 Non-Type Parameters for Templates

Besides the template arguements that are preceded by the class or typename keywords, which represent types, templates can also have regular typed parameters, similar to those found in functions. As an example, have a look at this class template that is used to contain sequences of elements:

Example 8-4: Non-type parameters for templates.

```
//-------------------------------------------------------------------
// File: example8_4.cpp
// This program defines a non-type parameters for a class template.
//-------------------------------------------------------------------
1    #include <iostream>
2    using namespace std;
3
4    template <class T, int N>
5    class mysequence {
6        T memblock [N];
7    public:
8        void setmember (int x, T value);
9        T getmember (int x);
10   };
11
12   template <class T, int N>
13   void mysequence<T, N>::setmember (int x, T value) {
```

```
14          memblock[x]=value;
15      }
16
17  template <class T, int N>
18  T mysequence<T,N>::getmember (int x) {
19          return memblock[x];
20      }
21
22  int main ()
23  {
24          mysequence <int, 5>    myints;
25          mysequence <double, 5>    myfloats;
26          myints.setmember (0,100);
27          myfloats.setmember (3,3.1416);
28          cout << myints.getmember(0) << '\n';
29          cout << myfloats.getmember(3) << '\n';
30          return 0;
31      }
```

Result:

1   100

2   3.1416

In this example, integer **N** is a non-type template parameter. When the objects of the template are created in Line24 and 25, an integer 5 is directly passed into **N**. It is also possible to set default values or types for class template parameters. For example, if the previous class template definition had been:

template <class T=char, int N=10>

class mysequence {..};

We could create objects using the default template parameters by declaring:

mysequence<> myseq;

which would be equivalent to:

mysequence<char,10> myseq;

8.5 Derivation and Class Templates

Class templates can inherit or be inherited from. For many purposes, there is nothing significantly different between the template and non-template scenarios. However, there is one important subtlety when deriving a class template from a base class referred to by a dependent name. Let's first look at the somewhat simpler case of nondependent base classes.

In a class template, a nondependent base class is one with a complete type that can be determined without knowing the template arguments. In other words, the name of this base is denoted using a nondependent name. For example:

```
1 template<typename T>
2 class Base {
3 public:
4     int basefield;
5     typedef int T;
6 };
7
8 class D1: public Base<Base<void> > {    // not a template case really
9 public:
10    void f() { basefield = 3; }         // usual access to inherited member
11 };
12
13 template<typename T>
14 class D2 : public Base<double> {       // nondependent base
15 public:
16    void f() { basefield = 7; }         // usual access to inherited member
17    T strange;         // T is Base<double>::T, not the template parameter!
18 };
```

Nondependent bases in templates behave very much like bases in ordinary nontemplate classes, but there is a slightly unfortunate surprise: When an unqualified name is looked up in the templated derivation, the nondependent bases are considered before the list of template parameters. This means that in the previous example, the member **strange** of the class template **D2** always has the type **T** corresponding to **Base<double>::T** (in other words, *int*). For example, the following function is not valid C++ (assuming the previous declarations):

```
void g (D2<int*>& d2, int* p)
{
    d2.strange = p; // ERROR: type mismatch!
}
```

This is counterintuitive and requires the writer of the derived template to be aware of names in the nondependent bases from which it derives—even when that derivation is indirect or the names are private. It would probably have been preferable to place template parameters in the scope of the entity they "templatize".

Templates and Inheritance
- Templates and inheritance relate in several ways:
  - A class template can be derived from a class-template specialization.
  - A class template can be derived from a non-template class.
  - A class-template specialization can be derived from a class-template specialization.
  - non-template class can be derived from a class-template specialization.

Think These Over

1. What is a template? Why is templates used?
2. What is the difference between function templates and template functions?
3. What is the meaning of template instantiation?
4. What is the difference between class templates and template classes?

8.6 Case Study: An Example of the Vector Class Template

You can use an array to store a collection of data such as int, float or user-defined type. There is a serious limitation – the array size is fixed when the array is declared. C++ provides a **vector** class, which is more flexiable than arrays. You can use a **vector** object just like an array. The vector size can increase automatically while an element is stored into the vector. Now we define a **Vector** class template and initialize it by using **student** class objects and **float** values.

Example 8-5: A **Vector** class with parameterized types.

```
//------------------------------------------------------------------
// File: example8-5.cpp
// This program defines the class template of a vector structure and initializes by using student class
// objects and float values.
//------------------------------------------------------------------
1   #include <iostream>
2   #include <cstring>
3
4   using std::cout;
5   using std::cin;
6   using std::endl;
7
8   //definition of the student class
9   class Student{
10  public:
11    Student(char *strname = "", char *id = " ", int g = 0)
12    {
13       name = new char[strlen(strname) + 1];
14       strcpy(name, strname);
15       ID = new char[strlen(id) + 1];
16       strcpy(ID, id);
17       grade = g;
```

```
18      }
19      Student(const Student& s)
20      {
21          name = new char[strlen(s.name) + 1];
22          strcpy(name, s.name);
23          ID = new char[strlen(s.ID) + 1];
24          strcpy(ID, s.ID);
25          grade = s.grade;
26      }
27      Student& operator =(const Student &s)
28      {
29          delete[] name;
30          name = new char[strlen(s.name) + 1];
31          strcpy(name, s.name);
32
33          delete[] ID;
34          ID = new char[strlen(s.ID) + 1];
35          strcpy(ID, s.ID);
36
37          grade = s.grade;
38          return *this;
39      }
40      operator int()
41      {   return -1000;         }
42      void print() const
43      {   cout << "Student: " << ID << " " << name << "; grade: " << grade << endl; }
44      ~Student()
45      {
46          delete[] name;
47          delete[] ID;
48      }
49   private:
50    char *name, *ID;
51    int grade;
52   };
53
54   //definition of a vector class template
55   template<typename T>
56   class Vector
```

```cpp
57  {
58  public:
59      Vector() :size(0), values(0), space(0){}
60      Vector(int len)                                    //constructor
61      {
62          size = 0;
63          space = len;
64          values = new T[len];
65      }
66      ~Vector()                                          //destructor
67      {   delete[]values;   }
68      Vector(const Vector& v)                            // copy constructor
69      {
70          size = v.size;
71          values = new T[size];
72          for (int i = 0; i <size; ++i)
73              values[i] = v.values[i];
74          space = v.space;
75      }
76
77      Vector& operator = (const Vector& v);              // overloaded operator =
78      T& operator [] (int index)  {    return values[index];    }
79      T at(int index);
80
81      template<typename T>
82      friend std::ostream& operator << (std::ostream&, Vector<T>&);   // output a vector
83
84      void push_back(T d);
85      void pop_back();
86
87      int getSize() const { return size; }               // get the size of the vector
88      void reserve(int newalloc);
89      void resize(int newsize);
90      int capacity() const {    return space;    }
91
92      bool empty() const {   return size == 0;   }
93      void clear() {    size = 0;    }
94  private:
95      T *values;              // the element of the vector
```

```
96     int size;                    // size of the vector
97     int space;
98  };
100
101 template <typename T>
102 Vector<T>& Vector<T>::operator=(const Vector<T>& v)
103 {
104     if (&v != this)
105     {
106      delete[] values;
107      values = new T[v.size];
108      for (int i = 0; i < v.size; i++)
109            values[i] = v.values[i];
110      space = v.space;
111     }
112     return *this;
113 }
114 template <typename T>
115 void Vector<T>::push_back(T d)
116 {
117     if (space == 0) reserve(8);
118     else if (space == size) reserve(2 * space);
119     values[size] = d;
120     ++size;
121 }
122 template <typename T>
123 void Vector<T>::reserve(int newalloc)
124 {
125 if (newalloc <= space) return;    //
126 T *p = new T[newalloc];
127 for (int i = 0; i < size; i++)
128     p[i] = values[i];
129 delete[] values;
130 values = p;
131 space = newalloc;
132 }
133 template <typename T>
134 void Vector<T>::resize(int newsize)
135 {
```

```
136    reserve(newsize);
137    for (int i = size; i <newsize; ++i) values[i] = 0;
138    size = newsize;
139    }
140    template <typename T>
141    void Vector<T>::pop_back()
142    {
143    if (empty())
144        cout << "Vector is empty.\n";
145    else
146    {    --size;    }
147    }
148
149    template <typename T>
150    bool Vector<T>::empty() const
151    {         return size == 0;         }
152
153    template <typename T>
154    T Vector<T>::at(int index)
155    {
156    if (index >= size)
157    {
158        cout << "Beyond boundary\n"; return values[index];
159    }
160    else
161        return values[index];
162    }
163
164    template <typename T>
165    std::ostream& operator <<(std::ostream& os, Vector<T>& v)
166    {
167    for (int i = 0; i < v.getSize(); ++i)
168        os << v.values[i] << " ";
169    return os;
170    }
171
172    int main()
173    {
174    int i;
```

```cpp
175    Vector<Student> sVector(3);
176    cout << "sVector size: " << sVector.getSize() << " sVector' space: " << sVector.capacity()
                      << endl;
177
178    char na[512], id[512];
179    int gr;
180    for (i = 0; i < sVector. capacity(); i++)
181    {
182         cin >> na >> id >> gr;
183         Student st(na, id, gr);
184         sVector.push_back(st);
185    }
186    for (i = 0; i < sVector.getSize(); i++)
187    {      sVector.at(i).print();       }
188    cout << "Appends elements at the back of the vector\n";
189    cout << "Remove the last element from the vector\n";
190    sVector.pop_back();
191    cout << "sVector size is " << sVector.getSize() << " sVector' space is " << sVector.capacity()
                      << endl;
192
193    Vector<float> fVector(4);    //declare a template class with float data type
194
195    for (i = 0; i < fVector.capacity(); i++)
196         fVector.push_back(float(i * 3 + 0.5));
197    cout << "fVector=" << fVector << endl;
198    cout << "Remove the last element from the vector\n";
199    fVector.pop_back();
200    cout << "fVector size is " << fVector.getSize() << " sVector' space is " << fVector.capacity()
                      << endl;
201    cout << fVector.at(2) << endl;
202    fVector.clear();
203    cout << "fVector size is " << fVector.getSize() << " fVector' space is " << fVector.capacity()
                      << endl;
204    cout << "fVector = " << fVector << endl;
205    return 0;
206    }
```

Input:

wang 13001201 78
gao 13001202 90

zhang 13001203 67

Result:

1. sVector size: 0 sVector' space: 3
2. Student: 13001201 wang; grade: 78
3. Student: 13001202 gao; grade: 90
4. Student: 13001203 zhang; grade: 67
5. Appends elements at the back of the vector
6. Remove the last element from the vector
7. sVector size is 2 sVector' space is 3
8. fVector=0.5 3.5 6.5 9.5
9. Remove the last element from the vector
10. fVector size is 3 sVector' space is 4
11. 6.5
12. fVector size is 0 fVector' space is 4
13. fVector =

Word Tips

template *n.* 模板	regular *adj.* 规则的
conveniently *adv.* 方便地，合宜地	indistinct *adj.* 不清楚的，模糊的
relate *as* 与……有关	instantiation *n.* 实例
stack *n.* 栈	encounter *vt./vi.* 遭遇 遇到
paste *vt.* 粘贴	automatically *adv.* 自动地
reinvent *vt.* 更改，重做	angle *n.* 角度
intelligent *adj.* 聪明的，智能的	appropriate *adj.* 适当的
reusability *n.* 复用性	specialization *n.* 专门化，特例化
arbitrary *adj.* 任意的，独裁的	retrieve *vt.* 检索，重新得到
parameterization *n.* 参数化	placeholder *n.* 占位符
generic *adj.* 普通的; *n* 泛型	generically *adv.* 一般地
vectors *n.* 向量	plug *n.* 插头
reuse *vt.* 重用	actual *adj.* 实际的 目前的
container *n.* 容器	double-stack *n.* 双栈结构
identical *adj.* 同一的，同样的	associate *vt.* 使联合
tedious *adj.* 单调的，乏味的	software-reusability 软件重用性
error-prone *adj.* 易错，易错的	last-in 后进
strategy *n.* 策略	first-out 先出
version *n.* 版本	non-template 非模板

Exercises

1. Find the error(s) in each of the following and explain how to correct it.

```
template<class T>
    T min(T,T);
    const int s=7;
    void k( )
    {   min (5,2);
        min ('a', 'b');
        min ('a',1);
        min (s,7);
        min (2.7,4);
    }
```
2. Define a *Swap* function template that implements exchange of two numbers. Implement the template by using *int* and *double*.
3. Define a **getMax** function template with finding the maximum value among 10 numbers. Implement it by us *int* and *float*.
4. Write a function template to compare two numbers. If the they are equal, then return 1; otherwise, return 0.
5. Define a class template **array** that produces a bounds-checked array. Implement it by using *int* and *float*.
6. Define a stack class template. Implement the template by using *int* and *char*.
7. Given the following program:
```
template <typename T>
class Vector {
public:
    Vector(int len);
    ~Vector();
    Vector (const Vector<T>& v);
    T operator ()(int i) const;
    T& operator()(int i);
    int getsize()                              // get the size of the vector
    Vector<T>& operator= (const Vector<T>& v);     // overloaded operator =
    friend ostream& operator << (ostream&, Vector<T>&);   // output a vector
private:
    T *values;          // the element of the vector
    int size;           // size of the vector
};
```
Write out the implementation of class template **Vector**.
8. Write a class template to search an element in an ordered array by using dichotomy.

References

[1] Bjarne Stroustrup. The C++ Programming Language (Special Edition). Pearson Press, 2001.

[2] Bruce Eckel. Thinking in C++. Prentice Hall Inc, 1995.

[3] Nell Dale, Chip Weems, Mark Haedington. Programming in C++. Jone and Barlett Publishers, 2001.

[4] D.S. Malik. C++ Programming from Problem Analysis to Program Design (Third Edition). Thomson Course Technology,2007.

[5] Harvey M. Deitel and Paul J. Deitel. The Complete C++ Training Course (Fourth Edition). Prentice Hall Inc,2003.

[6] Stanley B.Lippman, Josee Lajoic, Barbara E. Moo. C++Primer (4th Edition). 李师贤，等译. 人民邮电出版社,2006.

[7] Bjarne Stroustrup. C++程序设计原理与实践. 王刚,等译. 北京：机械工业出版社，2010.

[8] Walter Savitch. Problem Solving with C++: The Object of Programming (Fifth Edition). 周靖译. 北京：清华大学出版社. 2005.

[9] 钱能. C++程序设计教程. 北京：清华大学出版社，2005.

[10] http:// www.cplusplus.com/doc/tutorial/

内容简介

本书第 1 版被列入"普同高等院校'十二五'规划教材",自出版以来,在作者学校计算机科学与技术专业本科生教学已使用 2 年,得到了广大同学的赞许,而且也被其他院校用作程序设计初学者的双语教材。

本书在保持前版的特色基础上,对部分章节内容进行了修正和补充,增加了案例和练习题。书中在各个章节起始处提出学习的目标,重要知识点和易混淆知识点处均有重点注解和思考题,例题配有完整的程序代码和运行结果,每章末尾有重要的词汇注解和相应的练习题,有助于读者理解书中内容,帮助读者掌握面向对象编程方法。

本书共分 8 章,围绕面向对象程序设计中对象的作用,介绍了 C++进行面向对象程序设计的类定义和封装、继承、重载及多态等的实现过程和 C++中泛型程序设计模板的基础知识。

本书用通俗易懂的英语描述其内容,符合中国人的思维、语言习惯,既让初学者了解面向对象程序设计的原文表达,也便于对知识点的掌握。本书面向具有程序设计入门基础的读者,可作为高等院校计算机及相关专业的面向对象程序设计课程的双语教材。